D0211226

RIVERS IN THE DESERT

RIVERS IN THE DESERT

William Mulholland and the Inventing of Los Angeles

MARGARET LESLIE DAVIS

HarperCollins*Publishers*

HarperCollins books may be purchased for educational, business, or sales promotional use. For information, please write: Special Markets Department, HarperCollins Publishers, Inc., 10 East 53rd Street, New York, NY 10022.

Designed by Jessica Shatan

Library of Congress Cataloging-in-Publication Data

Davis, Margaret L.
Rivers in the desert : William Mulholland and the inventing of Los Angeles / by Margaret Leslie Davis.—1st ed.
p. cm.
Includes bibliographical references and index.
ISBN 0-06-016698-3 (cloth)
1. Mulholland, William. 1855–1935. 2. Water supply engineers—United States—Biography. 3. Water-supply—California—Los Angeles—History. I. Title.
TD140.M84D38 1993
979.4'93051—dc20 C.1 92-54718

---

93 94 95 96 97 ❖/HC 10 9 8 7 6 5 4 3 2

For Dr. Brian Young McLean

When nature has work to be done,
she creates a genius to do it.

Emerson

CONTENTS

Photographs follow page 146.

ACKNOWLEDGMENTS

William Mulholland's office files during his tenure as Chief Engineer and General Manager are part of the collections of the Los Angeles Department of Water and Power's Historical Records Program. The correspondence and other papers proved invaluable in chronicling the career of William Mulholland. I am especially indebted to Dr. Paul Soifer, project manager for the Historical Records Program, and consulting archivist Thomas Connors, both of the Bancroft Group, for their generous assistance.

Many people have made this book possible. I am deeply grateful to Joyce Purcell, Senior Librarian, Los Angeles Department of Water and Power, Craig G. St. Clair, company historian for the *Los Angeles Times*, Charles Johnson of the Ventura County Museum of History and Art, Kathy Barnes of the Eastern California Museum, and Thomas M. Coyle of the Los Angeles County Medical Association Library. I also thank the librarians at the University of California, Berkeley, Water Resources Center Archives, the Henry Huntington Library, the Moses H. Sherman Foundation, and the University of California, Los Angeles, Special Collections.

Special thanks to Jim Allen, Karen Chappelle, Michael Dougherty, Sandy Ferguson, Bob Feinberg, Jeffrey Forer, Louise Fraboni, Lee Harris, Burt Kennedy, Shelly Lowenkopf, Keith Lehrer, Steve Paymer, Roger S. Roney, Robin Shapiro, Rick Solomon, Richard Somers, Jean Stine, Paul Ward, David Williams, and Digby Wolfe.

Enough appreciation cannot be expressed to Catherine Davis and

James H. Davis, Noel Riley Fitch, Dr. James Ragan, Susan Vaughn, and the late Tommy Thompson.

My heartfelt thanks are due my editor, Scott Waxman, and literary agent, Richard Curtis. Finally, I am especially indebted to Larry Ashmead at HarperCollins, who graciously gave his support to this biography.

MARGARET LESLIE DAVIS
Brentwood, California
February 1993

RIVERS IN THE DESERT

PROLOGUE

It was July 26, 1935. Tens of thousands came to the Los Angeles City Hall to pay their respects to William Mulholland. Scores of black limousines circled the streets as mourners lined the sidewalks waiting to bid farewell to the retired chief engineer of the Los Angeles Department of Water and Power. At exactly 10:00 A.M., City Hall's huge bronze doors swung wide, and the waiting crowd streamed inside the building's massive, four-story rotunda.

Elaborate funeral wreaths of chrysanthemums, gladioli, and red and white roses surrounded the body, which lay in a flag-draped, blue steel coffin. Gifts, hand-written notes, personal tributes, framed photographs, mementos, and garden bouquets had been lovingly placed beside his funeral bier; a myrtle wreath from President

Franklin D. Roosevelt and a hand-penned note of condolence from former President Herbert Hoover were among the offerings.

City officials, celebrities, working men, and families silently queued toward Mulholland's coffin. Among the mourners were publisher Harry Chandler, banker Joseph Sartori, philanthropist George L. Eastman, humorist Will Rogers, University Chancellor Ernest C. Moore, six United States senators, four state governors, scientists, millionaires, engineering associates, and men who had worked with him in the tunnels and in the field.

As mourners passed the open casket, they stopped briefly to stare at the waxen features of "The Chief," now finally at peace in death. Some placed tokens and gifts inside his coffin or near the pyre. Others gently touched the brow of their beloved Chief, or whispered a prayer, then awkwardly moved on.

Meanwhile, throughout the city, eulogies praised William Mulholland for his honesty, modesty, valor, intellect, humanity, and, above all, his spectacular achievements for the city of Los Angeles. At 2:00 P.M., for ten minutes, two million residents of Los Angeles halted commerce to pay homage. Flags at all schools and public buildings were lowered to half-mast. Water in the Los Angeles Aqueduct was stopped for one minute as it flowed from the river in the Owens Valley. One thousand miles across the desert, ten thousand men working on the Colorado River Aqueduct paused with reverence to stand bareheaded, their steam shovels, drills, and tractors silenced in tribute.

"We are a forgetful generation," declared Los Angeles Mayor Frank L. Shaw, "but pray God that this community will never forget the everlasting debt of gratitude it owes this human diamond. His like we may never see again."

IT HAS BEEN SAID of heroes that for every devoted admirer won on the precarious climb to glory, two enemies are incurred. William Mulholland was no exception. Coupled with the outpouring of tribute was enough hatred, both within the city of Los Angeles and in

Owens Valley, 250 miles to the north, to prove he had labored and struggled in the world. There had been many among the mourning crowd who had come not to revere but to damn; some even blamed him for the violent deaths of their kin.

They, too, had left gifts among the tribute offerings. Placed inconspicuously amid the rose petals of a huge funeral wreath draped across the foot of Mulholland's casket was a small glass vial tied with a ragged fragment of red cotton, now faded and stained. Many would have recognized it as a commemorative from the opening of the Los Angeles Aqueduct. The tiny vial had been saved all those years by someone who had been in the crowd that day. When opened, the vial emitted the unmistakable, acrid odor of urine.

Yet even this dismaying commentary was lost amid the tributes and praise. Like Moses, William Mulholland had gone to the mountain and had brought back life—in the form of water—to a city dying of thirst. He transformed a land that could not support 250,000 souls into a flourishing oasis harboring millions.

ONCE THE PROCESSION at City Hall ended and the last bereaved were gone, the bronze doors of City Hall were closed. Now William Mulholland would be transported to his eternal place of rest, a mausoleum situated upon a sunlit rise overlooking the city for which he had accomplished and sacrificed so much.

Requiescat in pace great dreamer, great builder,
great friend of our fair and prospering city.

1

GENESIS

The good works of some are manifest beforehand.
1 TIM 5:25

WILLIAM MULHOLLAND and Fred Eaton set out for the Owens Valley from Los Angeles on September 4, 1904, in a two-horse buckboard. Their trek to Inyo County would take five grueling days, and the two friends decided to camp out along the way, living on a miner's diet of bacon, beans, and hard liquor. They later joked that their route could be easily traced by following the trail of empty whiskey bottles—dead soldiers—left in their wake.

The first twenty-five miles out of Los Angeles were uneventful. Eaton managed to negotiate the buckboard through the familiar dry washes of the Big and Little Tujunga Rivers without difficulty. The going got rough when they reached the notorious Newhall Grade where the narrow, unpaved road climbed forty-two degrees. The adventurers, one the chief of the Los Angeles Water Department and the other a former mayor, had to unload three weeks supply of food, water, horse feed, and bedding and push the buckboard behind the horses to get to the top, then trudge back down the grade to retrieve the supplies.

At the town of Newhall, in the Santa Clara Valley, Mulholland and Eaton spent the evening drinking at the local saloon. The next morning they traveled northwest to Saugus and east into the Soledad Canyon, where for thirty-five miles they struggled through a narrow and difficult mountain pass, and unexpectedly struck water. Their wagon sunk miserably into the soggy earth. The two civic dignitaries removed their boots and waded in, pushing, shoving, and cursing at the wagon and horses for two hours until the vehicle's wheels finally lifted to solid ground. After passing through the town of Acton—stopping at a small brick hotel where they liberally refreshed themselves at the bar—they journeyed fifteen more miles to the tiny, weather-beaten desert town of Palmdale, population twenty-five.

There, Mulholland guided the wagon across the summit of the Tehachapi mountains, altitude 3,800 feet, where he could see for a distance of 150 pollution-free miles the desert terrain that lay further ahead, a staggering vista of mountain peaks and dry lakes.

Entering the Mojave Desert after traveling through the junctions of Del Sur, Elizabeth Lake, Fairmont, and Willow Springs, Mulholland and his friend reached the sunbaked town of Mojave. After a night's drinking and rest in a deep featherbed at a Mojave hotel, the men departed shortly after sunrise. They had traveled a distance of ninety miles in two days.

Pushing forward across the baked desert floor, Mulholland heard the wheels of the buckboard crunch over miles of hard gravel and dry rocky washes. He saw the rock and boxwood headstones of men who had died in the desert, alongside the bleached skeletons of stage horses, their harnesses scattered along the isolated trails.

Some distance further, the desert melded into a gorgeous shade-mirage of turquoise and deep maroon, softening the hellish atmosphere of heat. When the wind blew, dust devils bounced wildly among the sage and greasewood; Joshua trees rose in the midst of nothing and stretched forth their twisted arms, as though warning travelers that the land's legacy was death.

The next day, Mulholland and Eaton reached a beautiful sandstone canyon called Red Rock, and came to the only sign of habitation within twenty miles—a shack of unpainted boards owned by an old Irish oxcart builder. The enterprising old desert rat had dug a water well and mounted a hand pump on it, and in front of the shack waiting for the travelers he had filled pails with water. A sign in lead pencil cautioned: WATER 10 CENTS A PAIL.

Drinking his fill, Mulholland joked, "Fred, there's no use traveling further—we've found the water for Los Angeles."

"The only problem," Eaton laughed, "is it's just too damn expensive."

Next, the two men reached the summit of the canyon, at last climbing to an altitude of 4,400 feet above sea level. As they progressed, the desert's floor rose higher and higher, in step with the peaks of the snow-capped Sierra Nevada mountains looming alongside them. After four days and nights, the two men approached Mt. Whitney. They ventured another twenty-two miles on the high plateau, crossing a number of small canyons. More than once, Mulholland and Eaton were forced to unload their packs and put their shoulders to the wheel to get the buckboard up the steep walls of rocky, dry creek beds. Finally, after passing through the desert town of Olancha, they reached their destination.

Standing bareheaded in the chill, William Mulholland beheld for the first time the breathtaking, spectacular body of water called the Owens River. Gleaming too brilliantly to look at directly in the morning sun, the vast expanse appeared in the distance like a great, silvered mirror. It meandered down the length of the valley where it discharged its waters into a large alkaline lake at the lower end. Bordered by lush salt grass, reeds, water birch, and willows, the banks of the roaring river were lined in red columbine, orchids, and tiger lilies.

Mulholland's engineering mind could not help calculating—even amidst all this beauty—that within the Owens River were flowing at least four hundred cubic feet of water per second, enough water to provide for a city not of two hundred thousand, but of two million people. The distance to Los Angeles was overwhelming, but Mulholland knew the Owens River sat at an elevation of four thousand feet, whereas Los Angeles lay only a few feet above sea level. The water, carried in open and closed aqueducts and siphons, could arrive at Los Angeles 250 miles south by power of gravity alone. As Eaton had told Mulholland earlier, costly pumping plants would not be necessary, and not one watt of pumping power would be required. Without a doubt, Eaton had discovered the resource that, once tapped, would free their city from stagnation.

"I thought you were crazy," Mulholland shouted to his friend over the noise of the rapidly moving water. "But our supply of water is indeed in the Owens Valley." It was one of the supreme moments of his life, and his thoughts would return to it many times, especially during his later years, after tragedy had struck.

William Mulholland had been taken to the top of the mountain; and like Moses he had been granted a vision of his people's deliverance. His miracle was not to part the sea, but to part the sands; not to keep the waters back, but to bring them forth and create rivers in the desert.

Mulholland and Eaton bathed in the cold, clear water, washing away the grime and sweat of their long journey. "Cleansing powers," Mulholland exuberantly called to his buck-naked friend, commenting on the water's purity, over the river's roar. "And with no equal." Mulholland, the former lumber camp stevedore, happily went unshaven, while the elegant Eaton peered at his reflection in the mirror-like water and shaved with a porcelain whisker brush.

In the midst of the vast Mojave, as coyotes yapped at the moon in the inky blackness, Mulholland and Eaton huddled in the flickering glow of their campfire in the shivering-cold desert night. They fondly talked about the Los Angeles which had so effectively shaped each of them and supported their careers, and about the vision they had for it once the precious water was delivered.

Although only two weeks apart in age, the relationship between the sophisticated Eaton and the unleavened Mulholland had initially been that of father and son. Eaton had groomed the industrious immigrant for promotion at the Los Angeles City Water Company, and Mulholland had paid dutiful respect and allegiance to his mentor.

Eaton's upper-class pedigree as the son of a prominent citizen and his dapper appearance contrasted sharply with Mulholland's poor Irish beginnings and his peasant's physique. While Mulholland's crude repertoire of ribald jokes at times embarrassed Eaton, he still appreciated Mulholland's disarming honesty and his simple love for literature and classical music.

In turn, Mulholland was intrigued that the smooth-mannered Eaton could guzzle hard whiskey in a smoke-filled poker game as easily as he could sip English tea at a political fundraiser. The friendship may have seemed curious, but it would prove to be the most important relationship between two men in Los Angeles's early history.

For the next ten days, Mulholland charted the valley and the river's course. His key problem was determining where the water could be diverted from the soda-filled Owens Lake, at a point before the river's water gathered, then wastefully evaporated—"doing nobody any good," except for the flocks of gathering lake birds that had adapted to its bitter salinity. He studied the problem in detail, tracing the proposed route of the aqueduct in improvised sketches and making rough surveys with an aneroid barometer and pocket level. Eaton and Mulholland calculated, measured, and debated every detail until they reached a rough agreement.

The region's beauty continued to fascinate Mulholland. The Owens River was bordered by lush salt-grass meadows, willow trees and cottonwoods. Reeds, rushes, aster, marigolds, desert buttercups, and floral colors of pink, lavender, white, and gold thrived along its banks. The river's waters ran the valley's full length, and were so clear, so pure, and so cold that they offered a haven for a wide variety of trout and wildlife. The Owens Lake brimmed with fowl, from swift-flying teal to honker geese, and as Mulholland and Eaton approached the lakeshore, thousands of them lifted en masse, taking flight.

Mulholland realized that without the Owens River running through the middle of the valley, fed by the eternally melting snows from the High Sierras, the area would be as arid as the Mojave Desert and its only life would be cactus, sagebrush, and chaparral. At the lower end of the valley, the river emptied into the 73,000-acre Owens Lake, an inland alkaline sea; its high soda content rendered it useless for irrigation.

At Owens Lake, Mulholland and Eaton camped overnight, cooking a goose in an open campfire, drinking whiskey, and smoking cigars. Mulholland lay face down in the luxuriant grasses, enjoying their coolness and velvety caress against his cheek.

"How wondrous are the works of the Almighty . . . and man is one of them," Mulholland murmured.

At daybreak, the two wanderers were off again. Mulholland maneuvered the buckboard through the placid valley. Eaton pointed to the abundant orchards loaded with peaches, pears, plums, and apples, and vines heavy with ripening grapes. Each ranch they passed straddled a stream from the rich river's waters. Their irrigated acres were loaded with bountiful crops of hay, alfalfa, and cereal grasses. First settled in 1861 by hearty pioneers who arrived in covered wagons, and later by successful livestock and mining companies, the lovely valley Mulholland was now exploring had grown into a network of farm communities.

When the first pioneers entered the valley, they settled along the river banks and dug irrigation ditches with hand tools, gradually diverting the river's adjoining streams of water onto the parched land, an acre or so at a time. For years the isolated, determined pioneers waged battle against heat, disease, famine, and floods. Slowly the desert bloomed with vegetation, and the canals were extended farther and farther from the river.

Finally, valley inhabitants constructed flood diversion canals to run down from the hills. Irrigation ditches traveled five miles or more from the river now to reach the secluded homesteaders. It was a water system that engineer Mulholland stopped to inspect and admire. Along the river, Mulholland also observed a series of small, prospering villages—Lone Pine, Independence, Big Pine, and Bishop. Unproductive land had been transformed into prosperous ranches; desert shacks had evolved into fine farmhouses, flanked by barns, silos, shade trees, and flowers. Settlers had built roads and schoolhouses. Now eight thousand people were living in the valley.

But all this wonder and bounty, wrought so tenaciously by the blistered hands of the valley natives, was virtually unknown to the far-off, troubled inhabitants of Los Angeles. Like conquering heroes, William Mulholland and Fred Eaton had discovered its beautiful secret, and as others would soon say, like thieves in the night, they were now conspiring to claim the valley's watery lifeblood as their own—no matter what the price.

IF MULHOLLAND'S BUMPY RIDE up to the valley from Los Angeles had been enlivened with liquor, his ride back was spent ruminating over more sobering thoughts. He realized that the problems of bringing water to Los Angeles would be immense, and the physical enterprise of the construction of an aqueduct would be staggering. It would be a momentous undertaking.

Though the Panama Canal, the New York Aqueduct, and the Erie Canal were larger and vastly more expensive, this project would be unique in water engineering because of its barren mountain and desert terrains. The Owens Valley Aqueduct would be the fourth-largest engineering project to date in American history, and the longest aqueduct in the Western Hemisphere. "It was as if Boston had decided to draw its water from the St. Lawrence River, or Washington, D.C., were reaching out to the Ohio, or St. Louis were reaching across the state of Illinois to Lake Michigan," author Kevin Starr would write years later, affixing the project's rightful place as one of the wonders of the budding twentieth century.

More important than the challenging engineering problems, Mulholland realized that such a monumental undertaking posed equally formidable political difficulties; the city council would have to approve it, though Mulholland believed they might now endorse any scheme that held promise. The thorny legal issues of water rights, city and federal approvals, and, naturally, sufficient capital would remain Mulholland's greatest obstacles. He braced himself for the countless problems that would have to be solved before construction could even begin. Surprisingly, the most formidable barrier to the project would turn out to be the one obstacle Mulholland never considered—and would never have thought possible.

Following Mulholland and Eaton's return to Los Angeles from their excursion in the Owens Valley, the two men embraced and said their good-byes. Eaton told Mulholland that he was traveling to San Francisco to visit his daughter. Mulholland assured Eaton that he would meet with the members of the Board of Water Commissioners and begin the battle to secure permission to build the great aqueduct.

Unbeknownst to Mulholland, Eaton boarded a train bound for New York. There he hoped to raise sufficient capital from investment bankers to secretly purchase the necessary water rights along the aqueduct route before the city had time to act. Eaton intended to sell the much-needed water to Los Angeles. The scheme, he calculated, could earn him estimated annual fees of $1.5 million. Fred Eaton's plan was to save the city of his birth and enrich himself immeasurably at the same time.

As Mulholland met in closed-door sessions with members of the Board of Water Commissioners, he was informed of Eaton's sudden ambition to gain control of the massive project. Mulholland was dumbfounded. Technically, there was nothing illegal in the proposal, but Eaton's apparent betrayal gave Mulholland pause. He viewed Eaton's deception as a personal assault and an egregious abandonment of the public trust. Mulholland feared the scheme would render the city's water supply hostage to the interests of private owners, and jeopardize construction of the mammoth project.

Mulholland's dream was of a vast, citizen-owned water and power system that would foster unlimited industrial and residential growth. For Mulholland, not profit but the unparalleled challenge of constructing the waterway and delivering the city from drought would be his enduring reward. For Eaton, the exploitation of Owens Valley water was an enterprise designed for financial gain.

Until now, Eaton and his protégé in the Water Department had maintained their close relationship. As a result of Eaton's plan, their twenty-five-year-long friendship would begin to unravel, and each man would come to view the other as his most dangerous adversary.

HAND OF BETRAYAL

Take heed
that ye not be deceived.
LUKE 21:8

THE BOARD OF WATER COMMISSIONERS and officials from the Water Department greeted Mulholland's revelation of the bountiful water supply in the Owens Valley with enthusiasm if not hosannas. But when they learned of Eaton's intentions to feather his own nest from the project at the city's expense, they were appalled that one of their own, a former city mayor, had decided to unfairly impede the city's progress.

Mulholland quickly contacted City Attorney and Water Department Chief Counsel William B. Mathews, who, by happenstance, was in New York City, and asked him to meet with Eaton and persuade him to call off his plan.

Mathews quickly learned it was too late. Eaton had already secured options on key tracts of Owens Valley land. Eaton announced that the Owens River water was now exclusively in his hands and that he intended to develop and control all hydroelectric power generated from the proposed aqueduct. Mathews relayed Eaton's grandiose ambitions to an outraged Mulholland.

To resolve the stalemate, Mulholland called upon a longtime colleague, Joseph Lippincott, a man who could change loyalties like a chameleon changes color, to act on the city's behalf and talk sense to Eaton. Lippincott had been in the Owens Valley as an official of the U.S. Reclamation Service, examining the feasibility there of a giant federal irrigation project. If the federal plan went forward, all the necessary land and water rights would be transferred from private to public ownership. Lippincott shrewdly told Eaton that the Reclamation Service would not withdraw from the Owens Valley unless the Los Angeles Aqueduct was "public-owned from one end to the other." With this news, Lippincott undercut Eaton's lofty dream of private water wealth.

Eaton quickly returned to the bargaining table with a new and more ominous scheme. Eaton had managed to obtain a valuable $450,000 option on cattleman Thomas B. Rickey's expansive Owens Valley ranch. To the wealthy and eccentric Rickey, the property—called "Long Valley"—was merely 25,000 acres of grazing land. To Eaton, it was the site for the only feasible reservoir in the valley, a level stretch of meadowland 20 square miles in size, with the potential to store approximately 183,000 acre feet of water. Eaton knew a massive storage reservoir at Long Valley would be critical to the long-term success of the aqueduct. In a transaction that should take its place alongside the purchase of Manhattan Island, Eaton only had to hand over a good-faith deposit of $100 cash to Rickey to bind the deal.

Eaton returned to Los Angeles and offered Mulholland a new compromise. Eaton would sell his Owens Valley land options and associated water rights to the city but insisted on keeping half of the Rickey lands for himself. He would retain 12,000 prime acres of the existing Rickey ranch (including the valuable Long Valley reservoir site) and extend a perpetual easement to the city for construction of a small reservoir at Long Valley. If the city did not accept his offer, Eaton told Mulholland, he would use the ranch as a haven for his budding cattle empire and take his offer elsewhere. There were others who would pay handsomely for an option on land sorely needed by the city of Los Angeles.

Eaton's threat triggered a wave of alarm in the halls of the Water Department where officials recognized the land's value. They would be forced to pay the price to the owner, whoever it might be, and better the devil they knew than some Eastern syndicate that might drive the price beyond reach. Eaton then added to tensions by speaking with reporters about his intentions in the Owens Valley and his comments made front-page news:

> Were it not for the fact that Los Angeles must accept the proposition as presented or lose all hope of saving itself from water famine, I would like to see the scheme defeated. This water right is a valuable thing and if it were not for the fact that the city would be robbed of what it needs most I would like to have them throw the lands back onto my hands.

Enraged at Eaton's self-serving public statements, and fearing the city council's ratification of the project was jeopardized, Mulholland clashed with Eaton in heated arguments for two straight days. The secrecy surrounding the aqueduct route was disintegrating, posing the threat of land speculation, and Mulholland was weary from long arguments and futile pleas; he finally reached a verbal settlement with Eaton in June 1905. The City of Los Angeles would accept Eaton's demands: In addition to his $450,000 price, Eaton received the 12,000-acre ranch, 4,000 head of cattle, horses, mules, and farm equipment—a sizeable return from his modest investment. He earned an additional $100,000 on commissions for other properties he secured on behalf of the city.

To celebrate, Eaton hosted an informal reception at the posh California Club, an exclusive men's club where Water Department officials and invited V.I.P.s met to congratulate him on his successful efforts in securing water rights for Los Angeles. Both Mulholland and W. B. Mathews were in attendance; despite the hair-splitting sessions they had been enmeshed in only days earlier, the three men drank heavily into the early hours of the morning and generously toasted the success of their future enterprise.

Eaton continued to ruminate over his lost multi-million-dollar dream. He felt his share of profits for solving the city's water problems was pitifully small and he was convinced he had needlessly sacrificed his financial ambition for an ungrateful city. "I have not received one dollar of city money, not even for expenses," Eaton complained to the press. Walking with a slight limp, his face pale and drawn, he appeared exhausted, and confided to friends that he was suffering from a severe attack of rheumatism brought on by his continued travels into the high altitudes of the Owens River country. "I have lived in this western country long enough to know what water is worth," Eaton blasted. "I probably know better than anyone else how much I could have made out of that option on the Rickey property. Private capital is waiting to put a million more into that valley than the city is buying it for."

Eaton worked himself into a frenzy decrying his loss of profits through civic fidelity. "Why, yesterday," Eaton moaned to the *Examiner*, "the old man Rickey came to scold me because I had thrown away a chance whereby we could have gone in together and made a big pile of money. He said I beat him and myself out of at least half a million."

Despite the warm reception of friends and colleagues who celebrated his civic beneficence, privately, Eaton was increasingly alienated. He was convinced he had been cheated out of a fortune and vowed to Mulholland that he would make no further concessions regarding the Rickey ranch.

Just days after the ebullient party at the California Club, Mulholland and Mathews re-examined their deal with Eaton, and soberly realized the city required a larger reservoir. They approached Eaton again, hoping to secure a more extensive easement in Long Valley which could be employed when the city grew large enough to require a permanent storage reservoir. Eaton told Mulholland and Mathews that he had given them "damn well enough for the money" and would not let them "flood his valley" under any circumstances.

Mulholland had intended to construct a 140-foot dam at Long Valley whose storage capacity would be double that of all other possible reservoir sites. Had he been able to erect a dam on the Long Valley site, Mulholland would have assured Los Angeles a supply sufficient to meet its needs through even the most prolonged droughts. Mulholland's plan would have preserved the Owens Valley as well and allotted residents enough water to keep 80,000 acres of first-class farmland under cultivation. It was Eaton's standoff over Long Valley which guaranteed, as William Kahrl wrote, "that there would be insufficient water for both Los Angeles and Owens Valley in any future drought, and g[ave] birth to the bitter Los Angeles Aqueduct controversy and the basis for the eventual sacrifice of Owens Valley."

The bargaining became so heated that Mulholland and Mathews left, threatening that "everything was off." But the next day they returned and obtained Eaton's begrudging consent to a reservoir only one hundred feet high.

As a consequence of Eaton's restriction, the city was forced to construct the needed reservoir elsewhere, a decision that would trigger far greater tragedy and ill-will than Eaton's clever machinations in 1905 could justify. To some, Mulholland committed a major error in failing to secure the Long Valley reservoir site, and he would later be accused of snubbing his nose at a more practical resolution with Eaton out of "petty niggardliness and almost fanatical pride."

The two men's deep-rooted feelings of mutual resentment were kept from public scrutiny. Despite the thorny contracts they hammered out, Mulholland and Eaton were forced by circumstance to unite publicly in their efforts to sell the project to the citizens and the press. In public they appeared cordial and warm although Eaton now maintained little enthusiasm for the Owens River project since he stood to profit less by it.

THOUGH FIFTY-ONE YEARS OLD, with hair nearly white (from two years of haggling in the Owens Valley, he claimed), Eaton was still remarkably handsome. In June 1906, fit and trim, suffering only from an occasional bout of rheumatism, Eaton quietly slipped into the city attorney's office at noon to marry his second wife, twenty-four-year-old city office stenographer Alice B. Slosson. Eaton intended to spend his honeymoon at Long Valley and designed a special heavy-duty motor car to traverse the sands of the desert and the little-traveled roads of the valley.

On hearing news of the marriage, Mulholland immediately sent Eaton a letter of congratulations, and recalled the day he confided to Eaton plans of his own impending marriage. Four years after he had succeeded Eaton as Chief Superintendent of the Los Angeles City Water Company, Mulholland married Lillie Ferguson, a fair-skinned, dark-haired native of Port Huron, Michigan. Lillie gave birth to their first child, Rose, one year later. She later bore him four more children: two daughters, Lucille and Ruth, and two sons, Perry and Thomas.

By year's end, Eaton had established a permanent residence at the Long Valley ranch, commuting to Los Angeles with his new bride once a month. His world had changed drastically. Once viewed by valley residents as their betrayer, he now seemed one of their own after his protracted bitter fight with the city. Soon this ex-mayor of Los Angeles would become the premier citizen of the Owens Valley.

3

SWEET STOLEN WATER

Ho, everyone that thirsteth,
come ye to the waters.
ISA. 55:1

AT A SECRET MEETING with members of the Board of Water Commissioners and leading department officials, newspaper owners were informed of Mulholland's visionary plan to build the longest aqueduct in the world. Fearing that Owens Valley land prices would skyrocket if news of the mammoth undertaking were publicized, publishers were sworn to secrecy in an unusual "gentleman's agreement."

To thwart speculation, the city had arranged to send Eaton back to the Owens Valley to acquire the necessary remaining options on downstream water rights below Long Valley. Furnished with official credentials (provided by the ever-agreeable Lippincott) which seemingly identified him as an agent of the federal government, Eaton met with unsuspecting Owens Valley farmers who thought they were aiding the proposed federal reclamation project as they signed over their water rights to the city, instead. On the same day that the Reclamation Service publicly announced that it was abandoning its project in favor of Los Angeles, Mulholland and Eaton returned to Los Angeles after a last buying spree in the Valley. "The last spike is driven," Mulholland happily announced to Water Department officers. "The options are all secured."

The city's deception at the expense of the unsuspecting Owens Valley landholders apparently was complete. Now the newspapers could make their simultaneous announcements of William Mulholland's plan for the great man-made river that would bring water to Los Angeles. Mulholland knew that the newspapers' editorials had to convince the people that the project was both necessary and urgent, so they would act quickly to approve millions of dollars in bonds and new tax assessments. Within twenty-four hours, the news would be public.

Unfortunately, the fourth estate's gentleman's agreement was breached. On the same day as Mulholland's announcement, an alert reporter in Independence wired the *Los Angeles Times* with the scoop of a lifetime. Fearing that the story might break first in the Owens Valley newspapers or leak from other sources, editors of the *Los Angeles Times* decided they dared not wait the remaining twenty-four hours before making their announcement. The following morning, *Times*'s headlines blared: "Titanic Project to Give City a River!"

Its front page revealed the city's whole plan, causing perplexed readers to consult their atlases to pinpoint the Owens River. Los Angeles readers responded to the news with acclamations of joy as they read about the "concrete river" which promised to increase their number to two million and transform the sun-baked San Fernando Valley into an agriculture-rich Eden. City water officials could not have written the story better.

"To put it mildly, the values of all San Fernando Valley lands will be doubled by the acquisition of this new water supply," stated the *Times* grandly but inaccurately. Within ten days of the story San Fernando Valley properties soared five hundred percent. Land options were gobbled up by competing realtors and new buildings sprouted over the arid land like mushrooms after a spring rain. Such was the force of the news that feelings of renewed wealth, prosperity, happiness, and fortune descended upon the city—even though nothing had happened yet.

Meanwhile, a number of rival Los Angeles newspapers (chief among them William Randolph Hearst's *Los Angeles Examiner*), which had agreed to hold back the aqueduct story until given the nod by water officials, smarted over the embarrassing scoop by the double-crossing *Times*. Hearst began a long series of diatribes against the *Times*'s editors, who responded in kind with heated accusations.

Insult led to insult, and soon a full-scale newspaper war raged. As vitriolic accusations were hurled from both sides and circulations skyrocketed, Angelenos followed the battling dailies with varying allegiances.

Meanwhile the people in the Owens Valley were outraged to learn not only of the federal government's desertion but of the predictions of their bleak future. "It probably means the wiping out of the town of Independence," stated the *Times* flatly, and a quote attributed to Mulholland that the land in the valley was "so poor that it didn't pay to irrigate it" added fuel to their mounting anger.

In Bishop the day the aqueduct story broke, Fred Eaton faced a menacing crowd of farmers enraged at what they considered their victimization by the former Los Angeles mayor. They threatened to string a noose around his neck. Barely escaping with his life, Eaton denied any wrongdoing, later announcing in an Owens Valley newspaper that he planned to spend his now considerable wealth and remainder of his life in the valley, adding, "in being a good neighbor I shall have an opportunity to retrieve myself and clear away all unhappy recollections." Then, once back in Los Angeles, he vented his rage upon water officials for allowing the *Times* to place him in such a life-threatening position. "They say I sold them out, sold them out and the government too; that I shall never take the water out of the valley; that when I go back for my cattle they will drown me in the river."

For his own dubious role in the scheme, Joseph Lippincott did not escape the rage of the duped valley citizens. His actions were criticized even in the Oval Office, where impassioned pleas were put before President Theodore Roosevelt by Owens Valley citizens and their congressional representatives to restore the Reclamation Service's original water project in the valley. As hostile feelings continued to grow among Owens Valley ranchers, area newspapers launched virulent attacks on Lippincott and "the Los Angeles cabal of water-seekers." Hatred for Lippincott, the man whom valley dwellers felt betrayed them most, ran so high that one angry crowd plotted to kidnap him, but at the last minute failed to carry out their plan.

In Washington, the Reclamation Service officials acted quickly to rid themselves of Lippincott's taint by demanding his removal from office. Lippincott quickly resigned and almost immediately was offered a more remunerative, $6,000-a-year job with the Los Angeles Board of Public Works. He would join Mulholland on the city's payroll as Assistant Chief Engineer of the Los Angeles Aqueduct.

On the Monday following the *Los Angeles Times* announcement, members of the Los Angeles City Council met to order an immediate polling on the issuance of $1.5 million in bonds to pay for land and water rights purchases in the Owens Valley. The first of two bond ballots (the first in 1905 for $1.5 million for land rights and the second in 1907 for $23 million in construction costs) was to be held on September 7, only three weeks away, notoriously the hottest time of the year. The council then adjourned to a champagne luncheon with William Mulholland, Fred Eaton, W. B. Mathews and the city's Board of Water Commissioners.

For the next three weeks, Eaton and Mulholland launched a vigorous round of campaigning, speech-making, and pressing the flesh. Night after night, they visited civic groups to urge their vote. Despite Eaton's waning enthusiasm as dreams of his potential windfall dwindled, his own interests forced him to continue his role as a leading project advocate.

Conveniently, the September heat delivered the final touch. As temperatures exceeded 100 degrees prior to the election and as water levels concurrently dipped, Mulholland used the city's oldest enemy—drought—to scare the daylights out of the hot and thirsty citizens.

In his basic stump speech, Mulholland wiped his sweating brow with his handkerchief and sympathized with the sweltering and tired members of his audience. He decried the "current emergency," warning that the entire city would soon be bone-dry. As the hot spell continued and water consumption skyrocketed, Mulholland refused to let the events go unheralded.

"This illustrates better than anything else could, the absolute necessity for securing a source of water supply elsewhere," he announced. "We must have it."

And have it they did. The Chamber of Commerce, the Municipal League, the Merchants and Manufacturers Association, and all leading civic and commercial organizations threw their zealous support behind the bond issue. Boosters heralded the gigantic undertaking as the key to creating a Garden of Eden in the southland, thus assuring Los Angeles's future prominence as one of the world's great cities.

Detractors pulverized the project as ill-conceived and ripe for graft, designed to make a handful of important men fabulously wealthy while bankrupting the citizens of Los Angeles with overtaxation. In one stinging charge that left the natives of Owens Valley feeling both irate and joyous, a respected Pasadena physician announced that his scientific tests revealed that Owens River water was a "vile bed of typhoid germs," and therefore, unfit to drink.

Mulholland, Eaton, and their entourage of assistants continued to feast and boost, sometimes challenging their endurance by dining at three or more banquets during the course of a single evening as they traveled from one appearance to the next. After delivering their speeches, they were applauded, interviewed, and photographed, and the two men would often conclude their remarks standing side by side with arms uplifted in victory, looking very much like a presidential ticket.

At one highly publicized appearance before the Municipal League at the stylish Westminster Hotel in downtown Los Angeles, William Mulholland was received with unrestrained enthusiasm. Prominent businessmen, members of the city council and judiciary, bankers, lawyers, and publishers—all the movers and shakers of the city with the means to get any issue off the ground—listened intently in concerned silence as he began a passionate appeal. By meeting's end, Mulholland had whipped the room into a frothy intensity, like a preacher conducting an old-fashioned revival meeting.

"If we could only make the people see the precarious condition in which Los Angeles stands!" pleaded Mulholland. "If we could only pound it in to them!" he added, pounding his own meaty fist.

"If Los Angeles runs out of water for one week the city within a year will not have a population of 100,000 people. A city quickly finds its level and that level is its water supply!"

Waves of wild applause erupted into bursts of cheers and whistles. Like the consummate stage performer he had become, Mulholland knew instinctively how to win his audience.

Historians disagree on how grave the "water famine" was that faced the city of Los Angeles in 1905. Some speculate that the drought was really a brilliantly orchestrated scare tactic. But internal memoranda from the Department of Water and Power and Mulholland's own correspondence indicate that prior to the election he had been forced by outright necessity to prohibit the sprinkling of lawns, restrict water flow into city fountains and parks, and order strict prohibitions of sewer flushing and domestic household water waste in order to preserve the water supply. He frantically installed elaborate pumping devices in underground artesian wells, and, when that failed, halted irrigation in the San Fernando Valley. Infuriated San Fernando farmers sued, causing Mulholland to spend countless hours at the county courthouse defending his department's actions.

Mulholland claimed reservoir levels during this period were the lowest he had ever seen. Weeks before the election, during the hottest part of the summer, the Buena Vista Reservoir's supply had fallen eight feet by 10 A.M. and sunk yet another foot before nightfall. Mulholland assessed the situation as critical and mandated a reduced water consumption of one million gallons a day. The city, Mulholland concluded solemnly, now "faced outright water famine."

Though he may have resorted to some exaggeration in his plea for votes, official records substantiate that Los Angeles's water supply had declined rapidly by 1905, and his concern for the city was genuine.

In the Owens Valley, local newspapers woefully predicted the valley's own demise and encouraged prominent citizens to fight the "water poachers" all the way to the White House if necessary. A bill was sponsored in Congress by wealthy anti-aqueduct landholders to prevent Los Angeles from using any Owens water for irrigation, but in a midnight meeting at the White House, President Theodore Roosevelt, who had signed the Reclamation Act to maximize use of natural resources, predictably vetoed the bill, giving Los Angeles permission to do with the water what it wanted.

GOSSIP AND INNUENDO surrounding Fred Eaton's financial interest in the aqueduct fizzled and by election day, he was virtually approbation-free. But shortly before the voters were to cast their ballots, new rumors surfaced that a powerful land syndicate had secretly bought up land options in the arid San Fernando Valley and stood to make millions when the aqueduct water arrived. Revelation of the syndicate's existence immediately threw the whole campaign into a tailspin, and posed the most blistering threat to passage of the bond issue.

Weeks before the election, the *Los Angeles Examiner*, still smarting over the *Times*'s scoop, broke a story some called the "Scandal of the Century" and claimed that through illicit communications long before the aqueduct's plan was made public, wealthy men were allowed to buy up San Fernando Valley lands at bargain prices.

The *Examiner* revealed that almost a year earlier, on November 28, 1904, less than three months from the day Mulholland and Eaton set out in their buckboard for the Owens Valley, a syndicate of private investors, acting on inside information supposedly limited to government officials, purchased a $50,000 option on the Porter Lands—16,200 arid acres in the north end of the uninhabitable and unfarmable San Fernando Valley—an option that, if the aqueduct were to be constructed, would be worth millions of dollars. The *Examiner* named ten syndicate members, each of whom held one thousand shares in the "San Fernando Mission Land Company," at a par value of $100 per share. The list included Leslie C. Brand, president of the Title Guarantee and Trust Company, rail magnate Henry E. Huntington, Edward H. Harriman of the Union Pacific, W. G. Kerchoff, president of Pacific Light and Power, and Joseph F. Sartori, president of the Security Trust Savings Bank. Of special interest to the tabloid writers were syndicate members Harrison Gray Otis, publisher of the *Times*; Edwin Earl of the *Express*; and Moses Hazeltine Sherman, trolley magnate and member of the Los Angeles Board of Water Commissioners.

The *Examiner* claimed that this syndicate, which had paid $35 an acre for the Porter Ranch, stood to make a profit of $5,546,000, as land values increased from $200 to $4,000 an acre. Although the option had been arranged before the syndicate learned the full details of the aqueduct's proposal, members of the group benefited handsomely from inside information provided by Moses H. Sherman who, as a member of the Los Angeles Board of Water Commissioners, could not have been a better set of eyes and ears. Apparently, he leaked Mulholland's intentions to build the giant system to his colleagues. Thanks to Sherman's dual role as Water Board Commissioner and syndicate investor, his cohorts were fully apprised of the city's plans in the Owens Valley, and were able to exercise their options on the Porter Ranch the same day that Fred Eaton telegraphed the water commission that his option on the Rickey ranch in Long Valley had been secured.

"Why should Mr. Eaton and his conferees have given the profitable tip to Messrs. Otis, Earl & Co.?" queried the *Examiner*. "Was this a consideration for newspaper support?"

The *Examiner*'s exposé caused members of the syndicate to nervously counterattack. The *Times*'s Otis accused publisher Hearst of misinforming the public, and smearing the names of prominent Los Angeles citizens. He called the *Examiner*'s Porter Ranch story "the very essence of absurdity. Mr. Otis and his fellows conceiving the Owens River project as a way to irrigate their San Fernando lands at public expense, has no foundation in the record." Equally incensed at the charges of his own duplicity in the affair, Fred Eaton tried to physically assault the *Examiner*'s editor by punching him in the nose.

Enemies of the aqueduct found a forum in the pages of the *Examiner* and promoted theories of corruption and conspiracy dating back to when "Old Bill Mulholland and his bunch of cohorts" created an artificial water famine to get the aqueduct project approved. The knockers protested that the giant project could never be completed for the $23 million Mulholland claimed and would collapse after the first earthquake.

Ironically, despite the plethora of reasons given by anti-aqueduct papers for voting against Mulholland's plan, their strongest argument was one they did not give: almost no one suggested the plan ought to be opposed because the water belonged to the residents of the Owens Valley. Los Angeles would be destroying its northern neighbor's livelihood in taking it.

THE SYNDICATE-CONTROLLED PAPERS portrayed Mulholland as a giant among men in carefully constructed stories, many penned by *Times* reporter Allen Kelly, who later left the *Times* to work in public relations for the water department. Otis and the syndicate recognized Mulholland's popularity and persuasiveness, and understood how much the public trusted him. By prominently featuring Mulholland in pro-aqueduct editorials, Otis and the syndicate members believed that the charges of graft and corruption against them might be dissipated.

Mulholland had the folk appeal of the rough-but-honest immigrant making his way among high-toned, slick business men. Born in Belfast in 1855, he had come West to seek his fortune. At the age of fourteen, Mulholland had left Ireland without his father's blessing to become an apprentice sailor. He crossed the Atlantic nineteen times during the next five years, visiting the ports of Europe, the United States and the West Indies, landing in Los Angeles in 1876.

Uneducated but strong and hearty with a quick mechanical mind, he found employment with the Los Angeles City Water Company, a private enterprise that had struggled to supply water to the nascent city of 15,000. His first job was as a *zanjero* or ditch tender for the primitive water works along the banks of the Los Angeles River. By the time the city acquired and municipalized the water company, Mulholland had replaced Fred Eaton as chief superintendent and Eaton had gone on to be elected city engineer and then mayor.

Mulholland had vigorously supported Eaton's candidacy. The two friends found themselves entrenched at the apex of city government: Eaton as mayor and Mulholland as water chief. City Hall was only a few blocks from the water company and Eaton and Mulholland met almost daily to drink and dine at their favorite downtown restaurant and talk political strategy. Nevertheless, Mulholland's image with the public was not that of the political insider, but rather that of the common working man.

News articles promoted Mulholland as a forthright character who "cut right through the bull." There was considerable truth in this image. Uncomfortable behind a desk, Mulholland spent most of his days in the field. He disliked writing reports of any kind and while other engineers took copious field notes and produced charts and reports, Mulholland relied instead on his prodigious memory. He left paperwork to subordinates; his business memos were invariably brief and he avoided correspondence whenever possible, stating that "if you leave a letter in the basket long enough, it will take care of itself."

In 1902, when the city sought to gain control of the private water company, Mulholland had been the company's chief officer during its long and bitter negotiations with the city. When Mulholland failed to produce the necessary records and inventories sought during the negotiations, many assumed that he was hiding information or acting vindictively. But as negotiations drew to a close, they discovered that these records did not exist—the company lacked even a simple blueprint of its existing water system. All data—down to the exact size of every foot of pipe, the age and location of every valve and pump—was carried in Mulholland's head.

The task of creating an engaging image for Mulholland was greatly facilitated for the press by Mulholland himself. He had a warm Irish accent, was quick with a witty turn of phrase, and was an entertaining public speaker. As election day approached, the land syndicate made effective use of Mulholland's personality. Their hero posed for photographers, gave interviews and attended political events, delivering rousing pro-aqueduct speeches. Making daily appearances at rallies, armed with charts, blueprints, and financial statements, Mulholland used his fiery oratory and considerable Irish wit to stir up voter enthusiasm: "If you don't get the water now, you'll never need it. The dead never get thirsty."

In later years, Mulholland supporters would recognize that in his zeal to promote passage of the bonds, Mulholland, in his paternalistic responsibility for Los Angeles and despite his own personal sacrifices, unwittingly allowed himself to be used by the syndicate for its own purposes.

Everywhere, the city was preparing for water. Pedestrians' lapels blossomed with buttons proclaiming, "I'm for Owens River Water." Attached to the buttons were small vials of the liquid. Passing automobiles were decked with banners, "Owens Water—Vote for It." School children paraded down city streets holding placards. Churches held special meetings, their congregations chanting "he showed me a river and everything shall live wheresoever the water comes." The Ladies Club hosted a week of special lunches—featured was tea made from Owens Valley water. The city's mayor named the Tuesday before the election "Aqueduct Day."

Mulholland continued his one-man show to the very end, visiting every meeting and rally he could. "Our population has doubled since

1904 while our water supply has diminished," Mulholland said at the final rally. "Owens is our only source, and defeat of these bonds would be fatal to the prosperity of the city."

On the morning of September 7, 1905, hundreds of cars and carriages hired by syndicate members and sponsored by special interest groups shuttled voters to the polls all day. Their labors paid off. After one of the most turbulent campaigns in the city's history, the bond issue passed overwhelmingly—by a margin of 10 to 1.

"Owens River is ours," a *Times* editorial exclaimed in celebration, "and our business now is to hustle and bring it here and make Los Angeles the garden spot of the earth and the home of millions of contented people."

Los Angeles voters had decided the future of what would become the principal metropolitan trading center on the West Coast. They turned to the man who would guide the entire mega-million-dollar undertaking to its conclusion. Mulholland now devoted himself entirely to the construction of one of the largest and most unique water systems in the world.

4

FAITHFUL SERVANT

Glorious is the fruit
of good labours.
SOL. 3:15

NOT SINCE ROMAN TIMES had such an immense water project been undertaken, and the Romans themselves no doubt would have been flabbergasted at the gargantuan project's details. The eighteen months of massive preparations, begun in the fall of 1907, would rival the excavation itself in magnitude and precision. Before work could begin, it was necessary to build roads and trails, power plants, telegraph and telephone lines, and to provide a water supply for camps established along 150 miles of waterless desert.

"The desolate barren sandy waste that for centuries upon centuries has felt only the light tread of the skulking coyote today crunches and creaks under the iron wheels of weary, heavily laden lumber freighters, and greasewood-bedecked steep canyons that had heard only the wailing cry of the solitary mountain lion resound with the tatter of thousands of hammers," described the *Examiner* as 500 miles of paved roads and rails, 240 miles of telephone wire, the world's largest municipal-owned cement factory, and more than 2,300 buildings (which included tent houses for workers, power plants, lumber mills, warehouses, barns, and hospitals) sprung up like magic over the vast terrains of mountains and desert.

Steam power could not be used in the desert because water was so scarce, so two hydroelectric plants near the Owens Valley intake, one at Cottonwood Creek and the other at Division Creek, were constructed to power the dredges, excavators, drills, shop machinery, and cement mills, and to light the camps and tunnels, making the aqueduct the first major engineering project in America constructed primarily using electric power.

As preparations began, Mulholland's overriding concern was not so much the immensity of the job, but rather whether bureaucratic officials would leave him alone long enough to do it. Principal excavation was scarcely under way in December 1908 when the powerful Los Angeles Chamber of Commerce, apparently ignoring the huge engineering preliminaries involved, anxiously demanded to know why so much time had been wasted and so little earth moved. Its chairman asked Mulholland for an informal report on the aqueduct's progress.

"Well, we have spent about $3 million all told," Mulholland answered tersely before the gathered members of the Chamber of Commerce, "and there is perhaps 900 feet of Aqueduct built. Figuring our expenses, it has cost us about $3,300 per foot so far." He paused as the shocked members gasped. Then he continued. "But by this time next year," he said, "I'll have fifty miles completed and the cost will drop to $30 a foot, if all of you here will let me alone."

The point was well taken. After a heavy silence, the chairman replied, "All right, Bill. Go ahead, do your job. We're not mad about it."

At first, Mulholland and water officials worried that the city would never be able to find enough skilled workers to construct the aqueduct, but a financial crisis in eastern markets had prompted the shutdown of dozens of mining companies in the western states just as major construction was about to get under way. By the summer of 1908, news of the Los Angeles Aqueduct had reached the beleaguered mining camps, and troops of transient labor forces from Nevada, Colorado, and Arizona traveled to Los Angeles to work on Mulholland's aqueduct. Newspapers fondly called them "blanket stiffs," a roisterous, whiskey-swilling bunch seasoned in the "drill and shovel work of great engineering achievements." Fresh from western mining colleges came a different but no less hardy breed—"eager young engineers who gained their first field experience in the rigorous desert life on Mulholland's Big Ditch and who proved their mettle as the backbone of aqueduct construction."

The living drama of five thousand men moving as one in the searing desert heat for the common good was a public relations bonanza for the city. Already, hopeful working families from eastern and midwestern states were packing up and heading to Los Angeles to reap the opportunities it would offer once the Owens water reached its arid boundaries. But the completion of the great aqueduct still lay years ahead. And the Chamber of Commerce extolled: "This is a public work without any politics. All employees are American. There is no contract labor employed. There are no men on the payrolls who have outlived their usefulness, or have been failures in life and have a berth because of friendship at city hall. Youth and virility fill the ranks of the 5,000."

In vivid portraits of words and photographs, rival newspapers issued progress reports and chronicled exciting derring-do details of life along the great aqueduct trail. The work, citizens were told, had no parallel in history, and day by day, week by week, month by month, year by year, these citizens would thrill and marvel at the heroic advance of their "Big Ditch." They, too, eagerly awaited the great arrival of the sparkling water that would irrigate 125,000 acres of land and harness 100,000 horsepower units of electrical energy, driving the city to unrealized heights of comfort and commerce.

Back in New York, publications like *Scribner's* and *Literary Digest* were also caught up in aqueduct fever. They sent their starched-collar correspondents to such Western locales as Jawbone, Grapevine, Mojave, Tehachapi, Olancha, and Haiwee to cable firsthand accounts of the antics of workers with names like "Goose," "Gunnysack Joe," and "Powder Face Kelly." These men, wrote the scribes, worked from dawn to dusk in the searing desert heat and bitter Mojave sandstorms, blasting, hammering, and drilling for their Chief and for the city.

From the start, William Mulholland's long experience and success supervising water projects in Los Angeles and elsewhere in southern California made him the Board of Water Commissioners' top choice to head up construction of the aqueduct. And it was no surprise, especially to the disgruntled citizens of the Owens Valley, that the newly appointed William Mulholland and the Board of Water Commissioners chose the congenial and accommodating Joseph B. Lippincott to be his assistant. A capable engineer, painstakingly patient with the paperwork that Mulholland abhorred, Lippincott was the ideal candidate. The Board also appointed former Los Angeles City Attorney W. B. Mathews as special counsel for the aqueduct and later for the Department of Water and Power. Long after the immense task had been completed, Mulholland would chortle, "I did the work, but Mathews kept me out of jail."

Water resources in Los Angeles were still severely depleted at the time of the aqueduct's ground-breaking, but thanks to Mulholland's previous attempts to conserve the precious liquid through metering and installing underground pumps, the present supply could meet consumption demands. When finished, the giant aqueduct would deliver 260 million gallons of water daily, and it was believed the city could continue to grow indefinitely without taxing the system's capacity.

The general plan of the aqueduct called for a straight cut of 22 miles of open canals, 43 miles of tunnels, 125 miles of steel siphons and concrete flumes, 137 miles of concrete-covered conduit, 13 miles of reservoirs, and a final 20 miles of connecting riveted steel main pipeline. This schema, as mapped out by Mulholland and the Advisory Board of Engineers, would be subject to such modifications and changes of location as might be found advisable during progress of the work. When finished, the route would stretch nearly 250 miles.

Thirty-five miles from the mouth of the Owens River, the aqueduct water would be diverted into an open canal, 50 feet wide and 10 feet deep, then for 20 miles it would travel above the river bed. Next, it would enter a concrete ditch, 18 feet wide and 15 feet deep, where it would be carried along the range of the Sierra foothills for 40 miles. The water then would pour into the 15-square-mile Haiwee Reservoir (where enough water was stored to satisfy Los Angeles's water demands for three years), and from Haiwee, it would travel a distance of 125 miles, through the desert via a series of closed tunnels and steel siphons laid into the rugged slopes of the western Mojave, where it would enter the Fairmont Reservoir, under the north slope of the Coast Range.

Released again, it would plunge into the mountainside, and for 5 miles course through the massive Elizabeth Tunnel, emerging on the southern slope of the mountains. At this point, it would tumble 800 feet onto the turbines of the two hydroelectric plants, enter another 7-mile-long conduit, then drop 700 feet to generate more power, pour through 16 miles of tubing to the wheels of a third power plant, and finally, come to rest in the San Fernando Reservoirs.

The Sierra Madre mountain range, which climbs to a height of 6,100 feet, was the greatest natural barrier to the construction of the aqueduct. The momentous task of conquering it would require considerable boring to create the Elizabeth Tunnel beneath a small, water-filled crater called Lake Hughes, 67 miles north of Los Angeles. The tunnel was the great connecting link to the whole aqueduct and by far the most difficult engineering challenge. But when completed, it would carry water from the Fairmont Reservoir in the Antelope Valley on the north side of the Sierra Madre to the San Francisquito Canyon. From there, the water would easily flow southward to Los Angeles.

Mulholland recognized that the time required to complete the entire aqueduct was in large measure dictated by the completion of the longest tunnel's construction, so he ordered that the north and south portals of the Elizabeth Tunnel be begun simultaneously in September 1907, at the same time preparational work began. Mulholland put two of his ablest men in charge to oversee the five-mile-long granite bore. The excavation was supervised by field engineer W. C. Aston at the tunnel's south end, and by experienced miner John Gray at the north portal.

A native of Pennsylvania, John Gray was hand-picked by Mulholland to attack the more difficult north portal because of his years of wet-tunneling experience in the mines of Colorado, Wyoming, and New Mexico. He was a good choice. Possessing limitless energy, he loved his work. He had hired his only son, Louis, to work as a tunneler in the Saugus Division, thirty miles south, and was proud of the work Louis was doing there.

Despite the inherent danger from falling rocks and dynamite blasts, the two Elizabeth Tunnel crews welcomed the work inside the excavation, where the temperature was a constant 58 degrees. Outside, the extreme desert conditions could range from a freezing-cold 10 degrees at midnight to a searing 120 degrees by midday. Other than the cement plant at Monolith where 250 men worked, the makeshift living structures in the camps were without insulation and did little to protect workers from the elements. One engineer recorded a temperature in excess of 130 degrees during the hottest part of the 1908 summer.

As Mulholland had hoped, by pitting the aggressive John Gray against Aston, he kicked off an intense rivalry between the men to be the first to reach the tunnel's center mark. A great game of nerve and skill developed between the north and south crews, placing them at additional risk in an already dangerous task. At first, the crews drilled powder holes for the dynamite charges with antiquated hand tools. Later, when the hydroelectric plants were completed, they were able to employ electric-powered air hammers which, although facilitating their work, were so loud that orders had to be given in hand signs. At the end of each day, every inch of excavation of the Elizabeth Tunnel was painstakingly measured by the crews. The strain of the toil was severe and the work painfully slow: if five feet were gained in a twelve-hour shift, it was considered a real achievement.

By mid-July 1908, the impatient and pressured Mulholland established a bonus system to hasten the snail-like pace at the Elizabeth Tunnel. A goal was set of eight feet per day; any crew which tunneled in excess of this rate would receive an extra forty cents per foot per man. Affording the men happy thoughts of extra pocket money, Gray and Aston organized highly efficient workshifts. Each progressed to nearly eleven feet of advancement per day. Any tunneler who could not keep up with the pace was given his walking papers.

The team tunnelers were a tough bunch, often violent and crude, and Gray handled their boisterous egos without favoritism. In turn, they respected his fairness, pluck, and dogged determination, knowing he would never ask them to go where he dared not go himself.

To be the first to reach the center mark in the Elizabeth Tunnel became an obsession for Gray, and despite the ever-present dangers posed by falling rocks and flooding, he drove his crews in a relentless pace to beat Aston's south portal crew. At times, Gray was able to double his gang's advance over the base goal. But he knew the going was generally easier at the south portal—there was less flooding and the bedrock was not as dense. This had allowed Aston to drive his men 604 feet forward in one month, setting an American record in hard-rock tunneling.

But such was Gray's zeal that during the period of Aston's record-making achievement, he worked through five straight shifts to try to catch up. He and his men ate their meals standing waist-deep in mud, sleeping only a few hours at a time, working round the clock to gain precious inches. Continuing to hit pockets of water, they were forced repeatedly to flee for their lives. They timbered the sides of the shaft, and drove overlapping steel rails ahead of the coring to hold back future possible cave-ins.

By sheer strength of will, Gray pushed his crews almost beyond their endurance, and with a little luck in the guise of an unexpected disaster, he managed to keep pace with the ever-advancing Aston. In

August 1909, Gray again struck deep water when his crew hit a large pocket of saturated sand and gravel. As one workman later described it, the tunnel shaft caved in with a loud, long "swoooosh," and thousands of gallons of water rushed into the tunnel. Work on the north portal was halted for forty-five days while Mulholland ordered an auxiliary shaft drilled into the tunnel at a spot 3,000 feet from the north portal entrance. This enabled Gray's men to attack both sides of the caved-in section simultaneously.

At first Gray threw up his hands in despair, thinking he had already lost the contest, but Mulholland's move actually gave him an advantage. The loose debris of the caved-in pocket merely had to be cleaned out by Gray's muckers while others of his crew gained additional footage by boring in from the north side. Despite continued flooding, the north portal crew was soon advancing almost as fast as Aston's. Relentlessly, Gray worked teams of between nine and twenty-five men in eight-hour shifts, eliminating downtime by developing a highly efficient relay system. As one crew of blasters completed its task, a new crew immediately arrived to begin work once the huge electric blowers cleared out toxic smoke and gases, then filled the tunnel with fresh air. Muckers would shovel the loose-blasted rock into waiting electric boxcars, while drillers set up dynamite blasts for the next team.

In the ensuing months, the steady sunset-to-sundown sound of distant muffled "pop-pop-popping" of the competing teams' charges was literally music to Mulholland's ears. Each "pop," he knew, signaled one more arduous step toward victory.

DR. RAYMOND TAYLOR was a familiar figure in the little northern towns of Lone Pine, Independence, and Bishop. His Franklin coupe with the alkali-stained water bag strapped around its hood ornament was frequently spotted rumbling along the gravel and dirt expanses of the aqueduct route. Born in Sycamore, Illinois, in 1872, Taylor had been hired by the city in 1907 to oversee the aqueduct's hospital system. He and his staff set up nine field hospitals at various points on the line. They recruited attending physicians and erected medical stores on hospital grounds. At each large construction camp, Dr. Taylor employed on-site medical stewards to ensure sanitary conditions and render first aid. When necessary, they would send severely injured and ill workers to the California Hospital in Los Angeles for treatment.

Taylor later said jokingly that he had joined the aqueduct to make money and for adventure—but he couldn't remember which was more important. Money wasn't Taylor's prime consideration—he relished life in the desert. Like Mulholland, he was an avid reader and spent most of his leisure time reading history books. Joining the aqueduct team as physician-in-charge not only offered him a chance to escape the boredom of city practice, but also gave him a once-in-a-lifetime opportunity to be a participant in what Taylor was sure would be one of the greatest historical engineering feats of the new century.

As work continued at a tumultuous pace, Dr. Taylor and his partners, Drs. Rea E. Smith and Edward C. Moore, examined nearly 5,000 men scattered over the 235 miles of aqueduct line. Each man had up to $1 a month in health care fees deducted from his pay. Meanwhile, Taylor and his partners' income soared. "We were beginning to make some good money. I had given myself a salary, about $500 a month, which was very nice for me." With it, Taylor was able to buy a two-story bungalow in Pasadena and eventually, as the project neared completion, his partners opted to retire to their comfortable suites at the elegant new Bradbury Building in downtown Los Angeles to handle paperwork. Only when emergencies warranted their presence would they travel to the line to aid Taylor.

The repeated trips up and down the growing length of the aqueduct route made Taylor an expert in desert driving techniques. The "four-banger Franklin" with its lightweight frame and underinflated tires could bounce its way over sand and rocks and through the dry sagebrush where heavier cars could not pass. Desert temperatures might soar to boiling point, but the Franklin's air-cooled engine would keep purring as long as Taylor stopped every so often to raise the hood and let the breeze cool it off.

Here in Owens Valley, Taylor witnessed the same natural grandeur that had smitten Fred Eaton and William Mulholland during their momentous buckboard trip of 1904. When duties allowed, Taylor visited Owens Lake, taking hundreds of photographs of the magnificent lake and its wildlife. Whenever he visited the aqueduct power plant at Cottonwood Creek at the foot of Owens Lake, Taylor sidetracked through the settlement of Olancha, an old stagecoach station named after the Shoshone tribe which lived south of Inyo County. Olancha's one-lane road was lined with tall cottonwood trees that cast cool shadows upon Taylor and his Franklin as they emerged from the desert sun.

It was always slow going navigating the sandy, rocky road to reach the power plant at Cottonwood, and repeatedly, Taylor would have to stop and re-start his car. By afternoon, his temper would grow short with the Franklin—and with the never-ending flies and the heat—and he would forget the beauty of the land. Cussing, he would wonder why he had ever left Los Angeles to go to work in such a godforsaken area.

It was at the Cottonwood Creek plant that Taylor had first met Harvey Van Norman. The young engineer was in charge of building the critical generating facility to supply power for the aqueduct. One afternoon, Taylor arrived to find Van Norman in a tirade. "The thing that was griping him," Taylor said later, "was that he had come up there to work on the promise that the officials would bring his wife up. He had been married only about three weeks." He had suffered uncomplainingly the delays and shortages that hampered his work but he was livid that he could get no word about his wife.

"If my wife isn't up here inside of a week I will quit and come on down," Van Norman told Taylor. Taylor had immediately liked the man's frank and open manner. He looked at the primitive conditions surrounding the Cottonwood Creek plant and felt sympathy for Van and his lost bride.

When Taylor left Cottonwood Creek later that afternoon, he had driven four miles in the Franklin when he saw a woman driven by motorcar coming up the steep grade. As they got closer, the two cars stopped. Taylor lowered his driving goggles and tipped his hat.

"Good afternoon, ma'am," he said warmly. "I know a man who's going to be pretty darn happy you're here." Bessie Van Norman smiled and Taylor saw the fine features and sparkle that had made Van a nervous wreck waiting for her arrival. Taylor noted she was wearing a wine-colored jacket with a pert hat and scarf to match. From her fresh appearance, he surmised that she had not been in the desert long, having made most of the trip by train.

"Terrible place for a woman," Taylor often said. "Two or three of them I know of nearly went crazy in the desert and had to leave entirely."

On his travels, Taylor would also take time out to watch the growing squads of workmen arrive on the standard-gauge railway that had been constructed to haul 320,000 tons of materials along the line. Taylor winced each time he read romantic portrayals of the workers in the newspapers. Behind the manufactured images of heroic crews laboring in efficient relays and surviving the dangers of hazardous operations was a different story, one with which Taylor was all too familiar.

The new men usually arrived on the job half-starved, tired, and very drunk. Many of them hadn't seen a square meal in weeks. Upon their arrival, they would sit down at the long tables of a crowded mess hall to gorge on what was probably the most food they'd seen in months. After a hasty meal, the men were introduced to a bathtub and told to scrub with lye soap and a washcloth. Their clothes and blankets were disinfected, and they were issued two clean bed sheets.

At each of the fifty-seven camps were portable wooden structures that housed drafting offices and cook shacks, double-roofed to fight the summer heat and "wire-guyed to resist the brutal cold northers." Alongside the portable structures stood rows of tented bunkhouses with dirt floors and no insulation. The bunkhouses were divided into rooms; sheets of rough cotton served as dividers.

Two men were assigned to each room; the lowest-paid workers, known as "crumbucks," cleaned these rooms for $2 a day. Sheet-iron stoves warmed the bunkhouses in winter and outdoor faucets from communal water reserves dribbled drinking water.

The camps were peopled with draftsmen, cooks, skinners, stake-punchers, chain-men, muckers, and various craftsmen who made up the construction crews. The typical lowly mucker was white, usually Irish, and around forty years old, although some were in their fifties and sixties. He wore woolen undershirts and drawers, dungarees, hobnailed shoes and, if he could afford it, a jaunty-looking derby that he would use as a hard hat. Though a falling tool would cave in the helmetlike derby, it might spare cracking a worker's skull.

With men's appetites to satisfy, wild saloon towns known as "rag camps" were built as close to the aqueduct line as safely possible. In the southern division, the largest of the boom towns, Mojave, became a mecca for ditchdiggers and muckers. On payday, the town roared, as though from a scene right out of an old Tom Mix movie. The town's streets were dotted with saloons, gambling joints, and whorehouses. Hordes of gritty workers crammed the tented saloons to drink, gamble, wench, play prairie pianos, and sing off-key with scratchy phonograph records. Fist-fights, knifings, shootings, and murders were commonplace.

One frustrated tunnel foreman admitted to Taylor that he constantly had to deal with "one crew drunk, one crew sobering up, and one crew working." To the consternation of alarmed city and water officials trying to keep the image of the aqueduct sparkling clean, Mulholland publicly admitted that it was whiskey more than anything that was powering the aqueduct's construction. "No man will do the hard, hazard-filled work of driving tunnels or skinning mules through canyons, while putting up with the blistering heat, biting cold, and dust storms without some relief," he said. Even Hugh Patrick II, Mulholland's teenaged nephew, was spotted in the taverns—where the proprietors graciously paid for his drinks.

Sadly, most aqueduct workers spent their money as fast as they earned it. "Today he is mucking in the tunnels of the Coast Range, recovering from yesterday's debauch in Los Angeles. Next week he will be at work in the ditch behind a steam shovel someplace on the desert. . . . The unknown always calls to him, and always he obeys the summons. Too parsimonious to purchase a pair of three dollar blankets or a pair of shoes, he spends a month's pay check with the munificence of a millionaire at the first point where he comes in contact with civilization and a saloon. He begins life anew with a raging headache and empty pockets," wrote Mulholland's secretary, Burt Heinley.

Stakemen, as the roaming miners were known, worked in as many as nine camps during three- or four-week periods; some gave different names each time. Each miner had his own goal, his own idea of what was called a stake, and the "short-staker" finished his tour of duty in a day or a week. Others would spend several months on the job, content with drinking their earnings in Mojave. And still others would work for several years and, after having saved a thousand dollars, return to Los Angeles, check into a hotel, and carouse until the money ran out. Some only reached the 18 Mile House, a bawdy saloon near Cinco. After a few days of severe insobriety, they would return to the line, unshaven and red-eyed, begging the crew chief to hire them back.

A large percentage of the aqueduct workers every summer packed up and departed for the other side of the Sierras or headed to the Pacific Northwest, where the weather was milder and the work easier. When temperatures cooled, they returned, drifting back on the trains bound for Mojave. As a result, Mulholland found that the winter months brought his most efficient labor. With that in mind, he rescheduled his construction plans to capitalize on the seasonal fluctuation.

As with any great enterprise, there were "human problems," as Dr. Taylor called the persistent troubles in the saloons. He knew the workers were "easy customers for sorry whiskey," and "it was a poor night when somebody wasn't knocked over the head, robbed, and thrown out in the back alley after he had gotten good and drunk. Then he would either bum his way out of town or return to the job." Men with bleeding heads were frequent sights at Dr. Taylor's office the night after payday, and the steward from the headquarters camp in Cinco would have to go down to Mojave three or four evenings a week to "sew somebody up or once-in-a-while pronounce somebody dead." The doctor was often paid for his work by saloonkeepers wishing to get rid of the "carcass" that lay crumpled on their floor.

Workers were able to obtain whatever medical care they needed, including hospital and surgical services, although Taylor and his fellow doctors were not contractually required to treat "venereal disease, intemperance, vicious habits, injuries received in fights, or chronic diseases acquired before employment." Laughingly, Taylor knew that if he and his partners were to take the contract's exceptions seriously they would have very little work left to do.

"I knew I didn't have to do it, but I just felt sorry for the poor cusses. They would work, many of them, until they got what they called a stake and then go to town and blow it on a big drunk if somebody didn't get it away from them the first night. They all love a fight, in fact, I think they get more fun out of scrapping than anything else, except getting drunk." But eventually, the saloon situation in Mojave grew so serious that even Taylor was leery about being on the streets at night. When required to sleep over, he stayed inside his hotel room, fearing to venture outside.

More than once Taylor lent twenty-five cents to a miserable, broke, hung-over hell-raiser so he could buy a can of stewed tomatoes. The tomatoes, eaten directly out of the can, were believed to be an antidote for drunkenness. "At first I thought the stewed tomato idea was pretty far-fetched and with no medical foundation," said Taylor. "I don't know why, but it certainly did work, and I always had a can or two in my bag to hand out." Not surprisingly, the demand for stewed tomatoes in Mojave grew so great that their price soared to over a dollar a can.

Fearing for the aqueduct's progress, the Board of Public Works eventually passed laws prohibiting saloons within four miles of any aqueduct construction. Thirty establishments were forced out of business. Proprietors challenged the law as unconstitutional, and the case went to the California Supreme Court, where it was upheld. After that, the saloons continued to thrive, but were constructed further away from the aqueduct's line. Whenever a new camp was established, a saloon inevitably opened nearby for business.

Taylor's partners openly disliked the outdoor life and gladly left the ugly side of field medicine in Taylor's capable, sympathetic hands. His city-slicker partners, with their Ivy League educations and lucrative medical practices, detested the primitive conditions in the field and the ailments associated with the rough laborers, in particular, alcoholism and venereal disease. But venereal disease was the least of the doctor's long-term concerns. What Taylor and company feared most was a sweeping epidemic of more common contagions that could pass from man to man, then jump from camp to camp. Not soon after his partners had departed, Taylor was confronted with his first full-scale epidemic.

Initially, he found that many men were repeatedly coming to him with painful, infected hands. At first, he was unable to make a diagnosis, but after examining dozens of men with similar open sores, Taylor concluded they were infected with impetigo, and that the men were passing the disease to one another with pick and shovel handles in contact with the open sores on their sweaty hands. Fearing that the disease would spread like wildfire hundreds of miles up and down the line, Taylor ordered all men with sores on their hands to immediately report for treatment. When the spread of the disease was not stayed, Taylor insisted that the pick handles and shovels be cleaned with antiseptic at least once a day. "This was hard to put across," Taylor wrote later. "The ordinary hard-rock miner is resistant to anything whatever that seems to look like an order, especially from someone they figured didn't know a pick axe from a shovel." Taylor was forced to enlist the help of some of the toughest men in the camps in his efforts to stop the impetigo. Eventually he succeeded in eradicating the disease.

For a short time, Taylor's duties returned to normal; he was making routine inspection trips up the aqueduct line twice a month. But one day, he discovered a worker at the Sand Canyon camp afflicted with typhoid. When others at the camp heard the word "typhoid," they approached Dr. Taylor with loaded shotguns and demanded that their co-worker be removed from their midst or be killed. But after examining the infected worker, Taylor realized that moving the patient to Los Angeles would kill him; the trip was too treacherous and grueling. First, the man would have to be transported on a cot in a horse-drawn wagon to a railroad station five miles away, then put into a baggage car and transported to Mojave. Then, after a six- to eight-hour delay, he would have to be transferred to a train bound for Los Angeles, then delivered by horsecart to the hospital.

The sick man was already racked with diarrhea; since outhouses were used at all the camps, Taylor speculated that the man may have already infected many other workers.

After placating the armed men, Taylor administered the "Brand Treatment" for typhoid, used routinely since 1861, consisting of a strict diet, nursing, and hydrotherapy. Since no ice was available, Taylor and his stewards soaked sheets in a tub of tap water, placed a rubber sheet underneath the patient, added a wet cotton sheet on top of him, then set up an electric fan nearby.

When the patient was finally stabilized, Taylor had the man transported to a Los Angeles hospital. He had succeeded in controlling a potential epidemic, and had allayed the workers' fears by urging some of the toughs who had helped combat the impetigo epidemic to support him in his efforts.

Exhausted, Taylor left Sand Canyon, and headed north to meet friends for a fishing expedition at Rae's Lake, near Independence. But on the way, while stopping at the Water Canyon camp, he diagnosed a driller with smallpox. The patient was put into the infirmary, isolated, and quarantined. Remembering well the lesson of Sand Canyon, Taylor placated the workers by offering free vaccinations. Trusting nature and luck more than any inoculation, only four out of three hundred men stepped forward to request the vaccination.

Ten days later, Joseph Lippincott arrived on the scene fuming with anger because Taylor had failed to quarantine the whole camp. "I had quite a time convincing J.B. that was a very foolish thing to do. I told him that the only quarantine that is effective in a camp is a shotgun quarantine with guards. The minute the men got word of a quarantine, they would attempt to leave, and shoot anyone who dared try to stop them . . . if there were latent cases of smallpox, it would tend to spread it all over the country as well as throughout our own camps." Despite Lippincott's apprehension, Taylor calmed the men, and prevented the spread of the disease. Taylor understood his rowdy and rugged patients and had a natural instinct for his unusual and sometimes dangerous medical practice.

5

NOISE OF MANY WATERS

We are laborers
together with God.
1 COR. 3:9

"JUST A MINUTE, NOW," John Gray chuckled deep inside the Elizabeth Tunnel, "and we'll light up our little Christmas tree." Covered in sticky clay like the rest of his crew, he activated the detonator, and nodded his head with each pop of the three exploding charges, counting each one as tons of three-billion-year-old granite and basalt were thrown into surprisingly well-ordered heaps, and clouds of black smoke and noxious gases billowed forward in the tunnel.

Once the big electric air blowers were moved into place and the smoke started clearing, the muckers, followed by Gray and the blasters, quickly returned to check how deep they had wounded "the beast"—the blasters' description of the granite core that they had been burrowing into for almost two years. Relaxing and smoking hand-rolled cigarettes, the blasters waited patiently for the mucking gang to remove the rock trimmings.

Suddenly, a massive cave-in occurred behind them, releasing tons of granite and debris. One and a half miles deep into the shaft, Gray and his crew shouted at one another in horror as a fifteen-foot-high torrent of water roared toward them. Scrambling for their lives, they clawed at the mounds of rubble, trying desperately to climb atop them to safety.

Thirty minutes later, a mucker, the first man to escape the deluge, staggered out of the mouth of the north portal. Ashen-faced and retching, the man told the workers who rushed to him that there had been a cave-in and Gray and the rest of the crew were trapped inside.

Immediately a rescue team was formed and entered the tunnel. Someone yelled to get hold of Mulholland and fetch a doctor. An old-time woolly tunneler, John O'Shea, now working as a crumbuck, moved to the mouth of the tunnel, dropped to his knees, and began to utter Hail Marys. After two hours, rescuers brought out four blasters dripping like wet dogs, bloody, and gasping for air. The rescuers went back into the tunnel, but the rescued men, bent over and heaving, said they were afraid Gray and the others had drowned.

At Jawbone Siphon, at Cinco, forty-five miles north, Dr. Taylor was informed of the cave-in. There had been other cave-ins at the Elizabeth, but none apparently as serious as this. Leaving behind the Franklin, Taylor quickly borrowed the Cinco superintendent's big, fast-moving Stoddard Dayton sedan and sped away.

By nightfall, Taylor arrived at the scene. There he saw William Mulholland pacing outside the tunnel, puffing nervously on his cigar. Taylor quickly walked over to him, shook his hand, and introduced himself.

"We're glad you're here, doctor," Mulholland said softly.

Taylor learned that Mulholland had been pacing in front of the tunnel for the past three hours. Grief-stricken, Mulholland told Taylor that it was like having his own sons trapped inside. Although Taylor had been working the line since commencement of the primary work, this was the first time he had actually met Mulholland, and was deeply impressed by his concern for the trapped men.

During the past ten years Taylor had heard and read so much about the legendary Mulholland and his amazing career that he was almost disappointed to see him in the flesh. The strain of the disaster, Taylor speculated, made Mulholland appear stoop-shouldered, and his frame smaller than he had imagined. Still, Taylor could barely conceal his excitement, hearing firsthand the thick, lilting Irish brogue that was described so often in the many articles and profiles he had read about the great engineer.

The two men, they would later discover, had much in common. Besides a love of the outdoors, both shared the vision of developing the city of Los Angeles through the creation of the aqueduct. When the bond issue came up for vote, the civic-minded Taylor was one of the first at the polls in his neighborhood, encouraging everyone he knew to vote yes.

As Mulholland continued to speak, occasionally veering the conversation to seemingly inconsequential subjects to relieve the emotional pressure, Taylor had no doubt that he was indeed facing the man who possessed all the sustaining qualities needed to complete the momentous undertaking. He knew that to the admiring citizens of Los Angeles, the Chief's personage bordered upon fable.

"The aqueduct's so damn big there's nothing in the ordinary mind to measure it by, unless you stand it up to old Bill Mulholland," proclaimed the *Times* for an admiring audience. Critics who wrote that "to try and tame the desert and attempt to build a concrete-steel river stretching hundreds of miles across the most unforgiving desert in North America was beyond self grandiosement—it was insanity," were squelched in editorials replying "Well, then, maybe he's crazy alright, but by God, he'll get the job done."

Studying the worried, tanned face with its field-marshal mustache, Taylor knew that much of Mulholland's success came not only from his dogged determination, but also from his remarkable ability to solve problems in the field. In his trips up and down the line, Taylor had listened to the many anecdotes of Mulholland's astounding memory told by admiring men who had witnessed it firsthand. Taylor remembered the articles written by *Times* reporter Allen Kelly, who covered news of the aqueduct. "I have been struck by the fact that he carries his data in his head rather than in note books. I never saw him make a note or refer to a note, yet at any point in the whole 233 miles he could show just where the line was located and tell the exact elevation as rapidly as a man can tell his own name and age." A statement by one of his engineers that "the Chief would rather get a shovelful of muck out of a tunnel than analyze all the cost data on the job" was widely repeated along the line.

Like a good general, he did not supervise, but directed and spent every waking moment he could next to his men, in the field. When faced with a thorny construction problem, he'd squat down and draw the solution in the dirt and move on, leaving others to work out the specifics. The Chief was, by his own account, "a practical man and a man of action who just liked to get things done." So far he had created a dedicated army of more than 3,500 diverse laborers which would eventually grow to more than 5,000. The army, armed with thirteen mammoth steam shovels, the shovels doing in a single day the work of 150 men, together with two dredges and a giant excavator, was advancing determinedly across 200 miles of desert. At the end of every twenty-four hours, the Chief and his aides without fail chalked up the aqueduct's progress, no matter how great or small, and set a new goal for the next day's labor. Remarkably, the excavation was progressing at four steady miles per month.

However, Taylor also knew of the conflicting perceptions of the Chief by the disgruntled among the ranks, who said that he was a harsh taskmaster, feared and admired alike, riding up and down the line like Genghis Khan on horseback, exhorting and driving his men on unmercifully, demanding that his engineers push the overworked brigades of dusty workers to do even better, and aim for five miles per month.

A story published in one tabloid underscored to many critics of the aqueduct Mulholland's obsession with keeping the project on time and on budget: when a tunnel worker, trapped in a cave-in, was kept alive by hard-boiled eggs rolled to him through an open pipe until he was rescued, Mulholland supposedly suggested dryly that the worker be charged board during the full time of his unplanned stay.

Now, listening to Mulholland quietly voicing his fears and hopes for the trapped men, Taylor found the accusations difficult to believe. For every negative story Taylor had heard, dozens more surfaced, telling of a caring leader who after dinner loved nothing more than to sit and joke with his crew crowded around him, relating ribald tales of his early adventures as a sailor and as a *zanjero* in early Los Angeles; of a leader who took a keen interest in the young engineers, encouraging them to bring their families to the line to live in the small cottages provided for them in the camps, knowing that a man who had a family to go to at night made a better worker and person. He enjoyed visiting them and immersing himself in the atmosphere of family life, roughhousing with children, assuming the role of godfather by handing out advice ranging from the correct method of changing diapers to the books that should be included in the children's education, and by constructing tented little red schoolhouses and providing teachers.

Whether tyrant or benevolent general, Taylor was aware the Chief's professional reputation was at stake and that upon his shoulders would rest the sole responsibility for the success of the enterprise. Any serious error of judgment would end his career, and the city's coffers would be emptied. The human challenge combined with the risk of utter failure made Mulholland's progress from now until the completion of the aqueduct the most important continuous news story in the West, and as such, Mulholland and his men were barraged with a constant trickle of newsmen and photographers covering every inch of the construction.

As the two men moved to camp chairs to sit, shouting was heard at the mouth of the tunnel. The workers and comrades of the trapped men were jumping up and down, shouting. John Gray and the last of his trapped crew emerged, alive and covered in muck.

Mulholland and Taylor rushed to them; Mulholland greeted his old friend with a slap on the back and embraced him. John O'Shea approached the imposing Mulholland. "It was the last Hail Mary that did it, Chief," he said, beaming a broad, toothless smile.

The spirituality of the moment was not lost on Taylor, as he recalled his readings of the prophet Muhammad. When he was asked what was the greatest act of charity a man could do, the son of the desert replied, "To bring water to men." Mulholland and his men, Taylor knew, were prepared to do so even at the risk of their own lives.

Gray, unharmed save for minor cuts on his face and hands, refused treatment from Dr. Taylor, and was back on the line the following morning at six o'clock, discussing with Mulholland his strategy for beating W. C. Aston and the south-end crew.

FEEDING THE ARMY of workers was a challenge that didn't go unheralded. The *Los Angeles Times* described the meal service:

> Three times a day the distant clank of the cook's triangle summoned the men to meals, and crews of tunnel stiffs, muckers, drillers and mechanics would stream happily and hungrily into the mess tents in answer to the call. Smiling eager young kitchen helpers, aspiring to enter the construction trade themselves, white starched aprons rolled around their waists, scurried from the kitchen with heavy steel pots of the choicest cut beef and plates of tasty viands.
>
> The fare for these hearty deserving souls was indeed fit for a King of the Realm. To start off their work day, they are greeted with breakfasts of steaming oatmeal mush, stewed fruits, thick sirloin steaks, scrambled eggs and bacon, fried potatoes, milk toast, hot biscuits, hot cakes, and coffee. After their taxing but rewarding day of labor, offered up for the choosing are rib sticking dinners of cream tomato soup, fricasseed veal, roast beef, baked beans, celery in cream, mashed potatoes, German salad, corn fritters, meat pie, cold ham, cold roast beef, combination salad, fried potatoes, spaghetti, stewed prunes, pudding, apricot pie, cake, cof-

fee and tea. Freshly starched blue linen table cloths and napkins complete the picture.

Eating at a ravenous pace, shouting good-naturedly, playfully swiping food off each other's plates, mischievously rapping one another on the head with soup spoons and ladles, letting off the pressure of the day, the happy din [of] their hearty comradery threatened to raise the tented roof of the eating hall. Full and contented, ready for a night of healthful slumber in their comfortable living quarters, the men left after a steward at the door marked their evening meal ticket.

In the engineer's mess hall section, where liquor was allowed, the evening meal was usually followed with a gentlemanly round of "rotgut-in-a-jug whiskey," chased with beer or soda. Here the evening typically ended with a visit to the camp's drafting room where the maps and blueprints were housed and discussions never wandered far from exciting talk of the enterprise that had brought them all to the desert—Bill Mulholland's "Big Ditch."

At first, as reported in the *Los Angeles Times*, the quality of food at the aqueduct camps was unquestionably superb, and food contractor Joe "D. J." Desmond knocked himself out putting his best foot forward. Besides the sumptuous meals, newspapermen and dignitaries found the dining halls to be clean and the kitchens free of pests.

Within weeks of obtaining his contract in 1908 with the Los Angeles Board of Public Works, which he won due to his skilled relief efforts in feeding thousands after the great San Francisco earthquake and fire of 1906, Desmond had established thirty-one cooking camps along the aqueduct line, charging each worker twenty-five cents a meal. He built three slaughterhouses with which to supply the kitchens, and set up canteens at each camp to sell tobacco, clothing, and other items.

Enthusiastic and conscientious, Desmond raced from camp to camp, overseeing his extensive enterprise in his chauffeur-driven limousine; but desert heat and lack of refrigeration made the food increasingly bad in proportion to the distance from Los Angeles, and the thirty-three-year-old youngest son of the founder of the Desmond clothing store chain had trouble living up to the standards of his exalted service.

In Desmond's defense, his driver Laurence Knapp years later summed up the problem:

> Joe was a hell of a guy. Maybe the meals didn't suit everybody, but they were always the best he could do. If you ever want to try to feed fresh meat to 5,000 men in the desert, with the temperatures from 100 up and no refrigeration, and men scattered over 200 miles, you just go ahead.

Meat spoiled, bread became infested, and most of the fare was restricted to simple imperishables. More than once, workers were driven to riot by the rotten grub. Tables were kicked over and food thrown on the floor, mess tents were torn down, and the cooks chased out of camp. And more than once, Desmond was warned by aqueduct officials that the business might be taken from him.

Compounding his own problem, Joe Desmond, apparently lacking in the basic subtleties of human nature, made the mistake of arriving at the camp kitchen at Sand Canyon in his shiny new black Mitchell limousine, and the men, already at the boiling point over the quality of the "sin-awful grub," whipped themselves into a frenzy of incrimination. When Dr. Taylor, just a few minutes later, arrived on his evening rounds, a full-scale riot over the food was taking place and the mess hall exploded before his very eyes. The men were busting the place up, throwing plates and bowls through the windows.

Desmond rolled down the window of the Mitchell and motioned for Taylor, who quickly got out of his Franklin automobile and ran over to Desmond and slid in the backseat of the Mitchell beside him.

"What the hell do they expect in Los Angeles?" Desmond said angrily. "These men are damn animals. Let 'em come up here and try and feed 'em!"

Taylor looked at the rioters in dismay and at Desmond in disgust. He knew he'd have to remain on the scene to stitch the men up after they had cooled down, but right now it would be foolhardy to try to stop them.

"You're pushing your luck," said Taylor.

Suddenly Taylor and Desmond heard shouting from behind them. They looked back to see the camp's shift boss, Big George Watson, accompanied by a group of men, running past them towards the mess hall, yelling at the rioters to stop. Inside the mess hall, Big George grabbed a metal-frame camp chair in each beefy hand and ordered the rioters to stop their "fussin' or else!" A six-foot-three, 240-pound tunneler, "hard as nails," Big George swung one of the chairs and knocked two men who advanced toward him to the ground, and then swung the other chair and knocked down two more.

"C'mon you sons o' bitches. I'll take care of you all," he growled.

In a panic the men jammed the doorway to get out of the mess hall as fast as they could.

Again Taylor looked at Desmond in disgust. In his mind's eye, Taylor was already counting the lacerations and broken bones he'd have to treat.

A few of the rioters rushed to the Mitchell and started pelting it with food, shouting obscenities.

Desmond ordered his driver to take off, and the limousine sped away. One mile from camp Desmond ordered his driver to stop. Desmond leaned back into the seat, exhausted, and Taylor started castigating Desmond about the food. For the past six months, the food had created an avalanche of protest up and down the line. In front of one camp eating house, a sign read DON'T MAKE FUN OF THE BUTTER—YOU'LL BE OLD AND SMELLY YOURSELF SOME DAY.

Desmond's ambitious efforts in the desert were in part an attempt to prove to his successful family that he too could turn a profit. But the inexperienced scion had failed to account for inflation in his contract, and by January 1910, Desmond's operation was in the red thousands of dollars. He then successfully renegotiated his contract to raise the price of meals to thirty cents.

Again turning a handsome profit, Desmond had purchased the

expensive Mitchell, a "practically perfect automobile." The big, powerful Mitchell could drive from Los Angeles to Mojave in two days, but the "perfect" axle broke four times. The big car was an eyesore in the primitive wilderness of the gritty desert, and to a short-staker sore about the grub, the very sight of the Mitchell was worthy of contempt.

Desmond's renewed financial stability, however, did not solve his problems. Without refrigeration, Desmond had to create makeshift techniques for cooling. He would slaughter beef in Mojave and then transport it after dark to the camps by automobile. It would take two full days to reach the most distant camps up the line, and by then of course, the meat was almost always spoiled. Without ice, meat and vegetables spoiled quickly. Weevils were found in the bread, worms in the sugar, and flies "encrusted in pies." The men complained about the "embalmed beef" and the "piss bitter coffee." One disgruntled worker presented to the *Los Angeles Herald* a photograph of bloated flies infesting an aqueduct livestock feed yard. Unable to resist, the *Herald* printed it in the Sunday edition, initiating unmerciful attacks by the press.

However, Fortune smiled on Desmond the profiteer. His slaughterhouses did not fall within the jurisdiction of interstate commerce, and so were never inspected by the federal government. The substandard quality of the beef continued, prompting a worker to comment that Desmond's unrefrigerated meat was "so full of holes from the maggots it looked like a lace curtain."

Two enterprising Frenchmen set up an eating house near Cinco that gave some of the men the opportunity to rejuvenate their poor stomachs. It was an instant success and when the men threatened they would buy their grub there, Desmond managed to convince aqueduct officials to issue orders prohibiting the workers from eating outside the camp. One group of workers retained Los Angeles attorney Olney S. Williams to file a lawsuit against the city over the quality of the aqueduct food, which stirred up considerable fuss and much publicity.

Mulholland steadfastly ignored the complaints about Desmond's food, believing that the city itself could scarcely do better in feeding thousands of men in the desert heat without ice or refrigeration. "The first three or four years of the work I boarded at the same camps with the men and never asked for a specially prepared meal," Mulholland informed the city council. "In all cases the average meals were good," adding that the widespread criticism directed at Desmond was unjust. "Our critics are those who haven't eaten three meals at the camp. . . . I never found Mr. Desmond unwilling to improve the mess when I suggested it," he said.

In the stormy course of his career, Mulholland was called many things, but never a gourmand. Although he knew the quality of the camp food would never change, he often, in the early years of the construction, complained the loudest about it to placate his men. But when it came to protecting the interests of his aqueduct, he was resolute. After his soul-hardening years as a seaman and working in the field, the food was more than adequate to Mulholland's taste. As long as he had his well-cooked meat and boiled potatoes, he was satisfied.

As a result of Mulholland's defense of the food, the situation went from bad to worse, and Desmond found himself the object of continued abuse from the workers, even from his own kitchen help. Cooking over coal ranges with temperatures rising to 120 degrees, flies everywhere, the conditions chipped away at the cooks' patience, with some walking off the job. Tempers flaring, the kitchen help often drank and fought among themselves; the hapless Dr. Taylor had to stitch them up and send them back to their bunk houses to sleep it off.

As significant as the food complaints seemed, they triggered a far more serious group of setbacks for the chief engineer. Within two weeks of the Desmond food price hikes, in November 1910, seven hundred men walked off the job at the Elizabeth Tunnel. Ever fearful of losing the edge he had over Aston to reach the center mark, when John Gray learned of the strike he ran to the camp office and rang up Mulholland, who was thirty miles away at the Jawbone division.

"The goddamn socialists are shutting us down!" he bellowed to

Mulholland on the telephone and implored him to do something. Hanging up, Gray ran back to the tunnel and frantically tried to enlist men to break the strike.

Mulholland arrived at the Elizabeth Tunnel the following day. Much to John Gray's dismay, Mulholland did not seem worried. Since construction had begun, the unions and now the WFM (Western Federation of Miners) had made frequent attempts to unionize the workers. Although some men had joined the union line, most of them up to now were reluctant to get involved in the labor movement. For the times, they earned top wages, especially those who competed in the bonus system, and according to one observer, the only legitimate complaint remained Desmond's food.

But this time when Mulholland drew a hard line and refused to discuss any pay increases to cover meal hikes, demanding that the workers return immediately, the metal workers and steam-shovel operators joined the WFM miners already on the picket line. After his gang of muckers joined the strikers in sympathy, John Gray stood in the tunnel alone, shoveling dirt and cussing. His only consolation was that Aston's team in the south portal had been shut down as well. Instead of sitting idly by, Gray stayed in the tunnel day after day, picking away at the granite.

The union men could not have picked a worse time to strike. In addition to the problems with labor, a shortage of aqueduct funds soon materialized. The eastern bonding firm which had been buying aqueduct bonds ahead of the established rate schedule decided not to buy additional bonds until the schedule and sale were even again. The labor troubles and a recession in the bond market had made investment bankers skittish, and New York bankers ceased purchasing aqueduct bonds altogether, creating a financial crunch that threatened to halt construction. Facing a catastrophic cash shortage, Mulholland traveled to New York to intervene.

Despite his increasing national fame, Mulholland was unable to persuade the New York money men, and frustrated, bitter, and tired, he returned to the line only to find that the WFM had now infiltrated the crews at the Little Lake and Grapevine tunnels and called an official strike.

Mulholland's brilliant organizational skill, which Taylor and so many others had admired, had finally worked to his disadvantage. It was a maddening dilemma—by pushing his crews faster and faster to the goals they were capable of achieving, Mulholland outdistanced the aqueduct funds available to him. If, on the other hand, he delayed work to stay within the schedule the financiers had dictated, Mulholland ran the risk that increased labor costs and inflation would drive prices way over budget and place him seriously behind schedule.

It was an utter disaster—the aqueduct was nearly two-thirds complete, and despite his pleas to the bond merchants, there was no money to finish. In this hour of desperation, Mulholland had only one course of action: He immediately announced that four thousand men, constituting 80 percent of the pipeline's work force, would be dismissed. "We're shutting down flat," Mulholland said sullenly. Overnight, his close-knit organization suddenly lay in shambles.

Mulholland now worked with a skeleton crew, suspending all work which was ahead of schedule so he could concentrate his remaining resources on the lagging sections of the aqueduct. At one point, fewer than 450 men were employed in the entire 233-mile project. Whole divisions along the line were closed down, with only the Elizabeth Tunnel remaining open to operate full-time. Without the boisterous workers, Mojave became a near-ghost town overnight.

Joe Desmond found himself stuck with huge losses, and as more workers joined the strikers, he was forced to raise meal prices a second time.

Raymond Taylor and his partners watched their income crash—from $5,000 to $450 a month. Taylor had sixteen bed-ridden patients in his field hospitals, no supplies, and no income. Fearful that the situation might drag on indefinitely, Taylor went to see Mulholland in the camp office at Jawbone, and told him that he was shutting down all hospitals but two, and intended to transfer the ill men to the infirmary near Mojave. Mulholland agreed. Taylor closed the hospitals at Tehachapi and at Haiwee, and accompanied the ill men to Mojave where he received a cautious telephone call from Mulholland's chief clerk.

"The Chief doesn't know when we'll open up again," the gloomy voice of the clerk informed him. "At least not until we get this money thing straightened out. And then when we start we'll have lost our organization and it'll take months to get it built up again. It'll probably never get as big as it was."

Taylor immediately contacted his partners, who agreed that they would be financially ruined if they continued to maintain their contract with the city. Taylor told them that he'd like to stick it out, reasoning that he couldn't be any worse off financially than if he were to quit. An arrangement was reached with the Board of Public Works: Taylor's partners were removed as participants, and the contract was assigned to Raymond Taylor as an individual. Taylor returned to the now-shattered line, in business by himself. Banking on a financial recovery, Taylor treated the remaining workers and engineers as needed, at his own expense, a tribute which Mulholland never forgot.

As the chief clerk had predicted, the work never attained its previous momentum. But in a strange, unexpected turn, the strike had provided much-needed relief from the current cash shortage for Mulholland. Without huge payrolls to meet, he was now in no hurry to settle the strike, and simply ignored it, biding his time until the money troubles blew over. For the strikers, Mulholland's move was as disastrous as their timing—their union quickly failed and they were all out of work. When the financial crunch eased in the spring of 1911, the New York bankers resumed buying bonds, allowing Mulholland to begin hiring a new work force.

At first, he found it difficult to enlist new non-union workers, and had to place full-page ads in the newspapers advertising for two thousand men to start the building again. Believing that the unadorned, direct words of William Mulholland would speak more eloquently than any from its own copywriters, the *Los Angeles Times* pleaded for workers on behalf of the city in a front-page ad: "There is work for any man who wants it. 'The Los Angeles Aqueduct needs men,' says Bill Mulholland."

"I was forced to go into the market and buy men as I bought lumber and cement and machinery," Mulholland would later state as he reflected on the crisis. To facilitate the induction of the new work force, he offered free transportation to the job sites, and a quick resumption of the bonus system. His only concern was that a new man accept his wages with no promise of an increase.

By May 1911, incoming crews were already hard at work. He had succeeded with only a modest delay.

AFTER ALMOST FOUR YEARS of single-minded effort, the two vying teams of the north and south portals of the Elizabeth Tunnel finally came face-to-face. On the previous afternoon, John Gray and W. C. Aston entered the screen door of Mulholland's makeshift office at the tunnel's camp and grinned at Mulholland hunched behind his desk pouring over blueprints.

"We're within spitting distance of each other," announced Gray.

"Give or take, at least that close, Chief," added Aston, grinning.

Mulholland smiled broadly. By his own calculations, the competing crews were now less than a few dozen feet from the center mark of the Elizabeth Tunnel. As good as the news was, he also knew a careless round of powder on either side would place both teams in imminent peril, and Mulholland cautioned the men to continue using the newly implemented tunnel communication system, a telephone line that ran over Elizabeth Mountain and into each portal, allowing the teams to communicate with one another at any given time. Now, at this close distance, communication between the two teams was vital for their safety.

When John Gray returned to the tunnel, he was agitated and nervous. Nightfall was quickly approaching, and he decided to rush one more shot even though his crew was quitting for the day. He drilled fifteen holes into the thin remaining granite bore and shot. As his muckers cleared the rubble, Gray felt the floor of the tunnel vibrate beneath his feet, and smiled. He knew Aston had just gotten off a shot of his own.

That night in his quarters, the hyperactive John Gray couldn't sleep and spent the night drinking coffee laced with whiskey, pacing and planning his tactics for the next day.

On the other side of the tunnel, Aston's crews huddled in their quarters around wood-burning stoves, placing bets among themselves which crew would be the first to break through.

"It'll happen by noon," Mulholland told his young assistant Harvey Van Norman before turning in. Big and broad-shouldered, engineer Van Norman's physical size was matched only by Mulholland's own.

"Who do you think'll do it, Chief?" asked Van Norman. "Gray or Aston?"

"It's even money," said Mulholland, taking a coin out of his pocket. "Heads Aston, tails Gray." He tossed the Indian head nickel high into the air. The coin landed on the floor and rolled into a crack between the rough pine planks.

"Well, I guess we'll just have to wait and see, won't we?" said Mulholland.

At dawn two terrific blasts were heard simultaneously from the south and north portals. The final hours of the great contest between man and nature was at hand. The blasts continued shattering the solid barrier of granite and gradually it was shaved down until less than nine feet of rock separated Gray and Aston.

Gray's and Aston's crews were working waist-deep in freezing-cold water, even as the pumps extracted 350 gallons a minute out of the portals. Covering the event, a *Herald* reporter, giving the citizens of Los Angeles a description well worth their nickel, wrote that conditions inside the tunnel were like a "titanic hollow serpent heaving in a cold sweat."

John Gray pinned his faith in the efficiency of his continued shots of dynamite. When the distance of the two sides was estimated at not more than four feet, Gray decided that two measured blasts would send his team through. On the south side, W. C. Aston had chosen to drill through rather than continue blasting, and the steady sound of his drills filled the south portal except for slight intervals to change drill heads.

Just before noon, Gray set off his last charge, estimating the crews would then be only about two feet apart, and he could then himself start drilling. But before the muck had been cleared away, Gray heard a noise scarcely louder than the pop of a small paper bag and saw the point of a Leyner Air Drill breaking through the thin granite wall, barely missing a mucker's head. Suddenly the wall shattered like a plate glass window, leaving a ragged opening twenty-one feet high and a scant six inches deep. Gray had overestimated the distance by a foot and a half.

There was a simultaneous, irrepressible cheer, and both crews crowded around the opening. Aston, grinning and sopping wet, was the first to step through the opening and into what was a second ago the north portal. "Hello, there, you north portal! Congratulations!" he roared with a smile as wide as the tunnel he had just conquered.

"Hello, yourself!" said Gray. "And same to you—lots of 'em!" and he moved forward to shake Aston's hand.

Mulholland, followed by Van Norman, appeared and threw his hat into the air and spit tobacco on the ground. "By God," he shouted. "She's done, boys. Well done!"

The 26,860-foot tunnel was excavated in 1,239 days, becoming the second-largest water tunnel in the nation. The concrete work needed to finish it took ten more months. The total cost of the tunnel was $1,611,600. The Board of Engineers had estimated it would take almost six years to completely finish the five-mile tunnel, and, including the cement work, the crews beat their deadline by more than twenty months at a cost savings in excess of $500,000, a considerable sum in 1911, and hastened the completion of the whole aqueduct. John Gray and W. C. Aston received unstinted praise from Mulholland, who later praised Gray's efficient relay system in engineering journals. Mulholland would soon send in another superintendent to supervise the tunnel's lining and pouring of concrete.

But for now the three men, intoxicated by victory, stood in the dripping, cold tunnel congratulating one another on a truly historic achievement. Overcome with emotion, John Gray sent for his son, Louis, who arrived the following morning. Father and son walked the entire length of the tunnel together, John Gray misty-eyed over the realization that the task was at last finished. Although the loser of the contest, Gray gained a lasting reputation as the best tunnel man on the aqueduct due to the difficulties he faced in the north portal. In their struggle, Gray's and Aston's crews repeatedly broke records for rapid hard-rock tunneling, and as progress on the Elizabeth Tunnel climbed to 22.1 feet per day (or a little better than 11 feet for each end), set a world record for hard-rock mining in May, 1910 with a single around-the-clock run of 567 feet, beating the Swiss at work on the Lichtberg Tunnel by 50 feet.

After a week of relaxation and heavy drinking, the two drilling teams disbanded, some moving on to other parts of the aqueduct system and others to different parts of the country. John Gray remained with Mulholland and worked until the opening of the aqueduct in 1913.

Upon completion of the Elizabeth Tunnel, the *Examiner* stated that "William Mulholland can sit back and twiddle his thumbs a little while, for the rest of the work is like child's play compared to this vast project." Accused of many things, it was one of the few times Hearst's yellow-sheeted tabloid could be accused of underestimating anything. The arduous tasks that still loomed ahead for Mulholland were far from a game of simple child's play.

6

BLOOD OF SACRIFICE

I have finished the work which
thou gavest me to do.
JOHN 19:28

DESPITE HIS PENCHANT FOR DARING, often dangerous ventures and his fondness for all things mechanical, Mulholland never learned to drive an automobile; he was transported up and down the aqueduct line in a chauffeured Water Department car. Usually he was driven from work site to work site by his second son, Thomas, or by his official driver. Despite the big, expensive company sedan and the ubiquitous driver, Mulholland—unlike Joe Desmond—promoted the figure of the working man. Stepping out of the sedan, his boots soiled by desert dust, Mulholland's demeanor was that of a man who knew hard labor firsthand. The men knew this and respected him; it would not have mattered if the boots had been patent leather. But behind the rugged, invincible image that Mulholland projected, the stress of the work was having its toll.

"This big project has completely worn you out," Taylor chided Mulholland, who was accompanying Taylor on an inspection of the newly reopened medical facilities. "The strain and responsibility is shattering your health. You have to slow down."

"What's your advice?" asked Mulholland wearily. "Am I to forget everything connected with the aqueduct? My work is nearly completed, and then I shall take a nice long rest," he sighed, knowing that over a year of conflict and turmoil lay ahead before he would be able to make good on this promise.

Taylor had developed his own afflictions from his work and the stress of insecure financial resources, among them a severe case of sciatica that made driving painful; but he bore his burden with good cheer.

On this day, Taylor observed again the early signs of what would later be diagnosed as Parkinson's disease. Mulholland's face was sun-scorched from his years in the field, and his heavy smoking habit contributed to its leathery, worn appearance. As the big, black sedan moved along, Taylor knew the slight twitching in Mulholland's face and the odd muscular jerking of the head would soon disappear after a few belts from the whiskey flask the two friends often shared on these trips.

Taylor knew that Mulholland's wife's illness was also adding to his stress. Lillie Mulholland, now confined to bed, was painfully suffering. Her condition had become so serious that Mulholland left strict instructions with the staff at the German Hospital in Los Angeles that he be notified immediately, anywhere in the field, if her condition should become worse. Lillie had never been very strong, and the doctors believed she was suffering from cancer. Every morning from his work site, Mulholland called the German Hospital for news about her condition. Despite his heavy workload, he made extra trips to Los Angeles to see her. Mulholland's eldest daughter, Rose, now twenty years old, faithfully monitored her mother's condition in near-daily reports to her father.

Mulholland was keenly protective of his wife and children. Especially as his wife's health deteriorated, he never burdened her with the pressures of public events or public attention and few of Mulholland's colleagues were aware of how seriously ill Mrs. Mulholland had become. Raymond Taylor, as both friend and physician, was one of only a handful of associates who knew the extent of her illness, and the toll it had taken on Mulholland's nerves. Due to the nature of her condition, and what little facts he could glean from Mulholland's silence, he feared that Mrs. Mulholland might die. It had to be a very difficult and shattering experience for the Chief; Mulholland was devoted to Lillie and he never hesitated to communicate with her by aqueduct telephone from the field or by telegram.

As the car moved along across the desert floor toward Mojave, Taylor produced the familiar, silver, half-pint whiskey flask from his coat pocket, unscrewed the cap, and passed it over to Mulholland. He smiled to himself as he watched Mulholland take a healthy swig and wipe his bushy mustache with his sleeve. Since that first meeting at the Elizabeth Tunnel cave-in, Taylor had never tired of being in Mulholland's company. He was pleased to see the tense facial muscles relax and the face appear younger than its fifty-seven years. Taylor realized that for Mulholland, not working would bring him down faster than the rigors of work, and his kindly advice to slow down would go unheeded. No matter what, Taylor knew that the Chief would continue his stupendous pace, making the arduous inspections of job sites and the many trips down to Los Angeles to give battle to whatever nay-sayer and bureaucrat happened to appear on the horizon.

Throughout the strain of his wife's illness and the rigors of the aqueduct, Mulholland had maintained a public image that was poised, refreshed, and confident. Now, in the warm glow of a good friend's company, the facade was no longer needed. "I don't know why I ever went into this job. I guess it was the Irish in me. Nature is the squarest fighter there is, and I wanted the fight. When I saw it staring me in the face I couldn't back away from it. I know the necessity better than any man; and if I don't, my thirty years of employment on the city's water works hasn't gone for much. I didn't want to have to buckle down and admit I was afraid of the thing, because I never have been—not for a second."

The strike and New York financial troubles that had threatened to destroy his spirit and the aqueduct itself were now behind him. The battle of the Elizabeth Tunnel had been fought and won. And Mulholland confidently told Taylor that he thought the worst was now over. If anything else came up, well, "Don't worry. We'll pull her through on time, never fear," he snorted, and passed the flask back to Taylor. Taylor, pleased at hearing the old optimism, never had any doubt that he would.

In spring of 1912, the whole aqueduct line was again in full swing. (Due to the extreme climate conditions, the number of workers was continually fluctuating. As many as 100,000 men toiled on the excavation over the period of its construction, with the work force usually numbering 5,000 at any one time.) The new crews continued in their favorite recreational pleasures and the Mojave streets were alive again and more decadent than ever. Back to making his rounds, Taylor looked upon all the familiar hell-raising and mayhem with mixed emotions. As a physician, he shook his head in despair at the broken noses, bloody fists, and cracked ribs. As sole owner of a profitable business, he could only smile.

IN OCTOBER OF 1912, despite the fact that the aqueduct was progressing beyond his own exacting expectations, Mulholland had to contend with what many considered his thorniest problem in the mammoth undertaking—Fred Eaton.

Making the long, hot trip down by buckboard from his Long Valley ranch to Mulholland's headquarters at Mojave, Eaton had watched with dismay the advance of the aqueduct as it stretched out before him for miles in a great line of open canals, steel pipes, and concrete conduits. The excavation sites of the future catchment reservoirs, eventually to be filled with Owens River water, scarred the terrain as far as Eaton could see, while workers by the hundreds toiled in the burning desert landscape. To anyone else, it would have been a rare sight of a great human endeavor, but to Eaton it was a blatant theft of his great enterprise. For five years, he had followed the aqueduct's progress with mounting rage, and at every reported foot of advance, every conduit finished, and every tunnel bored, he knew another acre of property was purchased in the San Fernando Valley by those he was sure had deprived him of his own rightful profits.

Since his bitter compromise with the city, Eaton had spent the ensuing years regretting its outcome in dark periods of wishful recriminations and self-imposed isolation, harboring feelings of injustice that friends reported were eating away at his pride. At the diversion site of the aqueduct above Independence, he had stopped to drink and fill his canteen with the Owens River water that was already trickling into the gray tufa concrete intake canal and had spit it out in disgust, its cool, sweet taste galling him as if it were the bitter water from briny Owens Lake.

Now as his buckboard slowly made its way toward Mojave following the line of the aqueduct, the events of the compromise played over and over again in his mind like a tormenting melody, and by the time he walked into Mulholland's office he was seething with anger.

Greeting Eaton with outstretched hand, Mulholland's face soon hardened as he prepared himself to listen to yet another long-winded harangue. Twice before, Eaton had visited Mulholland at the aqueduct and each time had demanded that Mulholland, on behalf of the city, meet his price of one million dollars for Long Valley, which he had decided to sell outright. Biding his time until Mulholland's labor problems had been settled, Eaton now chose to take up his bargaining again.

Although the planned reservoirs—at Haiwee, Fairmont, Bouquet Canyon, Dry Canyon, and two in San Fernando Valley, from which the water entered the Los Angeles distribution system—were in various stages of completion, Eaton knew these reservoirs functioned basically as catchment devices in the aqueduct's downward flow. The city, he also knew, would have to have a long-term storage facility for times of shortage and drought. Eaton's Long Valley, a twenty-square-mile meadow at the head of the Owens River, was a natural site for such a reservoir. Eventually, he knew the city would have to utilize it. It made no sense whatsoever to build the aqueduct without a storage facility at the source.

Recognizing his old friend's increasingly hostile behavior in each repeated visit, Mulholland knew that the situation required kid gloves, and despite Eaton's invective, Mulholland patiently explained the position he and the city held regarding Long Valley. Had Los Angeles acquired the property early during the construction phase when Eaton first made his offer, Mulholland might have been able to persuade the city to return to the proposal he had made in his 1907 report: a simple dam across Long Valley, 165 feet high, 525 feet across, creating a reservoir of 26,000 acre-feet that armed Los Angeles against dry years. But during their negotiations Eaton was not willing to give up that much, allowing rights for only a 100-foot dam, which was totally inadequate for the job.

Since then, after construction began, the Board of Public Works had eliminated it in its overall downscaling to save on expenses, and decided upon a permanent reservoir at another site rather than Long Valley, a site yet to be determined. Eaton's continuing demands for a million dollars only convinced the board that building a dam at Long Valley was not feasible.

Now, once again, Mulholland explained the city's position, and once again Eaton stormed out of Mulholland's office, slamming the screen door behind him. Flushed with anger and muttering to himself, Eaton strode to the buckboard and headed back to his prized Long Valley ranch. Eaton's visit, like the others, left Mulholland feeling depressed and anxious. There was no solution to his onetime mentor's anguish. His friendship with Fred Eaton seemed irreparably damaged.

On the heels of Fred Eaton's troubling visit, Mulholland was soon faced with the problem of defending himself back home in Los Angeles.

IN NOVEMBER 1911, Socialist candidate for mayor, Job Harriman, supported by the very trade unions Mulholland had defeated, made the repudiation of the aqueduct the pillar of a furious campaign. Charges of conspiracy and fraud leveled against Mulholland by the Socialists and other factions had been so loudly publicized that after much debate the Aqueduct Investigation Board, appointed by the Los Angeles City Council, began an official inquiry into their allegations, and in the summer of 1912 summoned Mulholland to Los Angeles.

Mulholland welcomed the investigation, hoping it would clear the charge of malfeasance on his part, but grew impatient when the inquiry turned into a political quagmire of "insinuation and crap," as he bluntly summarized it. He had answered charges in four long days of straightforward testimony about every conceivable detail concerning the purity of the water, the labor force, and the aqueduct's design—but he refused to cooperate with the board's repeated questions about Fred Eaton. Mulholland emphatically stated it was "unnecessary and unethical" for an investigative board to delve into the personal lives of the men who had conceived the idea of developing the aqueduct.

"You are not willing to recognize the right of this board to decide for itself the ethics of the case?" asked one examiner.

"I am the guardian of my ethics," Mulholland shot back, rising to his full, six-foot height. "It would be unethical for me to disclose names of men who never had anything to do with this thing. All this was prior to any deal in which the city is concerned. . . . This board has nothing to do with that," he continued, his fatigued eyes addressing every member.

Disgusted with the political motivation for the proceeding, Mulholland erupted before storming out, "The concrete of the aqueduct will last as long as the pyramids of Egypt or the Parthenon of Athens, long after Job Harriman is elected mayor of Los Angeles." The following morning, the quote was printed on the front page of every newspaper in California.

In spite of Mulholland's fiery defense, the investigation board recommended suing Mulholland, Fred Eaton, and special counsel W. B. Mathews, claiming they had deceived the electorate in an elaborate scheme involving water dumping and land fraud. The board's final report criticized nearly every aspect of the aqueduct, alleging Mulholland and the San Fernando land syndicate had artificially created the drought of 1905 in the minds of Los Angeles voters to induce passage of the aqueduct bonds and that Mulholland and the syndicate secretly plotted to hatch huge profits in the San Fernando Valley once the water had arrived. The report also alleged kickbacks to food contractor Joe Desmond and irregularities in Raymond Taylor's medical services.

The report concluded that the Investigation Board had insufficient time to "unearth the graft" which they were sure was hidden somewhere within the department and in Mulholland's administration of the aqueduct, but wanted to "hang the dirty wash" out on the line for the public to see for themselves.

Mulholland was not without powerful supporters, however; George Alexander, then in city hall as mayor, called the report "the veriest rot," claiming there was no graft whatsoever to uncover. As all of his friends knew, William Mulholland loathed the political game. To him, the inquiry was a sham, mired in political haranguing between pro-socialist and pro-capitalist forces, the proceedings swinging from critical attacks against the syndicate's influence in the San Fernando Valley to petty indictments that dark blue rather than white enameled tableware was used in the mess halls.

After four days of grueling interrogation, the exasperated Mulholland was almost persuaded to quit. "No independent man who fights for what he thinks can succeed," he told Raymond Taylor, who was appearing as one of the accused. "This is no democratic process," he announced. "This is my goddamn ditch."

The Board lacked power of subpoena, and could not legally require the appearance of the members of the syndicate, including Harrison Gray Otis, Moses Sherman, and Harry Chandler. Two other members were now out of the picture: E. H. Harriman of the

Southern Pacific rail system fortune was dead, and Henry Huntington, having sold off his local railway empire in 1910, was now in comfortable retirement, devoted to building his renowned San Marino library.

Although the syndicate's culpability was not proved, the proceedings were embarrassing, and scrutiny of the land speculation publicized the immense profits the members stood to garner once water was delivered to the valley. The entire affair proved to be an ongoing public relations nightmare. In the end, however, the inquiry did not demonstrate any tangible evidence of illegal activities and the board could only recommend that no water be provided to the San Fernando Valley for agricultural irrigation.

To Mulholland's satisfaction and vindication, on the central issue of municipal corruption, the inquiry asserted "there was not a statement submitted nor any evidence unearthed to indicate that any criticism of the integrity, business ability, or loyalty of effort for the best interest of the city of Los Angeles could be maintained to the slightest degree," concluding that "no direct evidence of graft had been developed."

However, to Mulholland's sorrow, the board called for the immediate indictment of Fred Eaton over his purchase of Long Valley. No criminal charges or civil litigation was commenced, and Eaton continued to sit on his Long Valley cattle ranch waiting impatiently for his day in the sun. The sense of loyalty to Fred Eaton that was born in the muddy water of the *Zanja Madre* was strong enough for Mulholland to defend him against such charges, even if he knew them to have merit.

The shoddy hearing over, Mulholland, with a sense of personal vindication, returned to the desert and to the task of bringing the aqueduct to its conclusion, only to be faced again with the most heartbreaking of all his many problems.

In June 1912, en route to Mojave to treat men injured in a barroom brawl, Taylor received word of a huge explosion at the Saugus division. Forty minutes later, he arrived at the south portal of the

Clearwater Tunnel, fifteen miles north of Saugus, to find three men dead and a dozen others buried alive beneath tons of blasted rock. Taylor's worst fears had been realized.

When he got there, two of the badly burned corpses had been recovered. Another twenty men had escaped, but were seriously injured by flying rock and the subsequent stampede by the workers themselves through the black, smoke-filled tunnel. "It was a scene of chaos, ugliness and tragedy," he reported later.

Shift boss Norman Stoble had been setting the powder for the next charge, and the others had walked nearly three hundred feet toward the mouth of the tunnel, when the powder suddenly exploded, blowing Stoble to bits. The blast blew in the width and roof of the tunnel, trapping the crew inside. An hour later, a second cave-in occurred, and tons of falling rock narrowly missed engulfing the rescue gang as they worked feverishly with shovels and picks to free the men trapped in the original blast.

As Taylor entered the tunnel, he saw the recovered bodies of Edward Garside and Thomas O'Donnell and the escaped men prostrate on the floor of the tunnel. The rescuers were attacking the huge mound of shattered rock and were shoveling debris into box cars while shouting out the trapped men's names in encouragement. Faint, muffled shouts from behind the mound were the only indication that anyone was still alive. Taylor immediately began treating the men who had escaped. They had been impaired by the bitter fumes of nitroglycerine as they fought their way through the wall of blasted rock to the open air a half-mile away. Some of them had become disoriented in the pitch black tunnel, running hundreds of yards in the wrong direction.

Six unconscious men were recovered and, with the help of two hospital stewards, Taylor succeeded in resuscitating each man. He learned that the tunnel foreman was still inside. When Taylor realized that the man they were talking about was Louis Gray, the twenty-seven-year-old son of John Gray, he dispatched one of the men to phone Gray at the Elizabeth Tunnel and bring him to Saugus at once.

Hearing that his son was buried alive and that three men were dead, John Gray fell to his knees in prayer. Informed of the accident, Mulholland dispatched a car to drive Gray to Clearwater Canyon. Gray arrived and tried to find out exactly what happened.

"The way these fellows handled dynamite and caps would make your blood run cold," Taylor told him, shaking his head.

Some speculated that the blast may have been caused by a candle igniting one of the fuses, because, in spite of the fact that the tunnel was lit electrically, the older miners insisted on working with candles—even though their use was strictly prohibited. A second theory suggested that the explosion was triggered by the "careless crimping of a cap on one of the charges as the fuse and fulminate primer was applied." Ironically, five minutes before the blast, a worried Louis Gray had emphatically warned the men to take more care with the explosives.

Gray ran into the tunnel; joining the rescuers, he frantically began shoveling rock and dirt, shouting his son's name to keep up his spirits. By midnight, the rescuers had broken through to the trapped men. Mulholland was relieved to learn that none of the buried men were badly injured, but were suffering from severe shock. An exhilarated John Gray emerged from the tunnel hugging his son. The younger Gray could not speak, and John Gray feared his son had been rendered mute by the explosion. Later his fears were relieved when Taylor pronounced Louis's condition as only temporary. He regained hearing in one ear, and Taylor later sent him to Los Angeles for treatment.

Mulholland worked vigorously to safeguard his crews, but by the very nature of their dangerous jobs, accidents occurred often. "It seems a miracle that only five men died during the whole of the construction," he stated later, referring to the tunnelers whose jobs were the most risky of all. Compared to the large number of deaths that occurred during the completion of the New York Catskill Aqueduct where 160 men died—nearly one man every week—experts applauding the minimal loss of life said credit was due in part to Mulholland's insistence on safety, including the use of expensive, modern, German-made fuses which reduced the risk of unintentional detonations.

When death did occur, the hard-bitten crews, heads bowed, hats held to their breasts, stood around the makeshift graves, wet-eyed at the loss of one of their own, and it was their Chief who, fighting back his own tears, delivered the simple, elegant eulogies.

Stoble, Garside, and O'Donnell died before they reached the age of thirty. Stoble's identification card stated that in the event of an accident his sister, Ann Stoble, of Redruch Highway, Cornwall, England, be notified. O'Donnell had a wife, Helen, living in San Bernardino. Mulholland determined that Norman Stoble's body was unfit to be transported to his family overseas. Instead, preparations for a "proper Christian burial" were made in the desert.

The city of Los Angeles would later settle death claims for the deceased men. The average amount paid per death claim was $436.20. Many of the roaming, highly independent, and often suspect workers refused to give out the names of next of kin and often no living relatives could be traced.

With Gray and his son standing at his side, Dr. Taylor fought back the tears hearing Mulholland's simple, plain words as Stoble, Garside, and O'Donnell were lowered into the hard, desert earth. Mulholland stopped work at the Coldwater camp site for forty-eight hours, feeling "tribute was more important than the damn work schedule."

From the outset the biggest challenge of the construction had rested upon the brave shoulders of these tunnelers. When finished, their crews would have bored through mountains of granite and sandstone, welded 100 miles of iron conduit, run 2,400 mule teams and exploded 6,000,000 pounds of dynamite to complete the Herculean task—a great chain of 164 water tunnels through mountains of granite, a feat that captured the attention of the engineering world.

THE LONG, DANGEROUS DAYS of building the aqueduct, however, were not without their lighter moments.

In October of 1912, the tunnels above Saugus were finished. The vibrant purple, white, and yellow hues of the verbena and primrose blanketing the desert floor had given way to the monotonous grays and browns of fall. In the high desert elevations, the air at night was already bone-chilling, and Dr. Taylor thought it would be the ideal time to give his young son Richard a tour of the aqueduct before winter set in.

At Dry Canyon, Mulholland, standing before the entrance of the newly opened tunnel, insisted that Taylor drive through it in his Franklin.

"Go on right through, " Mulholland said smiling.

"Can you do that?" Taylor asked cautiously.

"Sure, I've already gone through it two or three times myself. It'll save you fifteen miles, and only take a few minutes with your car."

Pleased, Taylor and sixteen-year-old Richard, who would later become a doctor like his father, started into the mouth of the tunnel in the Franklin. A small shallow stream of water was running out of the tunnel and Taylor, concerned, shouted back at Mulholland.

"Don't worry," Mulholland yelled back. "We're just running a little water to settle the ground so we can finish up the concrete work."

Taylor pulled ahead, switching on his lights. He saw there was just enough room to drive the Franklin in one direction through the dark tunnel. The smooth surface of the tunnel floor made an excellent driving surface and the Franklin moved along at a good clip until the car hit a section saturated with water. Taylor downshifted and the Franklin plodded through the next section of the tunnel where the water was even deeper and the floor was thick with mud.

The Franklin slowed to a crawl. Richard, who was having a great deal of fun and enjoying it all, jumped out and started pushing the car. The wheels sank deeper into the mud, but continued to inch forward. They had traveled about a mile into the dark depth of the tunnel when suddenly the Franklin came to an abrupt stop.

"Damn it!" Taylor shouted, looking under the hood. "There's water in the magneto."

The Franklin had no battery, and was powered by Bosch magnetos which when wet would cause the car to stall. Taylor and Richard looked alarmed. The headlights were still on, but they couldn't see more than a dozen yards ahead in the pitch black tunnel.

Icy cold water was dripping from the ceiling onto their necks.

"Jesus Christ," Taylor sighed and looked at his son in dismay. Taylor pulled off the magneto, wiped it dry and put it back on again. To their relief, the Franklin started up and they moved forward. The Franklin stalled again, and again and again; each time the exasperated Taylor got out and had to dry off the magneto to start up again.

Over an hour later, the Franklin managed to come out of the tunnel into daylight.

"It took us longer to get through that goddamn tunnel than it would if we'd gone all the way around by road," Taylor later complained to a bemused Mulholland. "I thought you said you did it all the time."

"Yup, that's what I said," replied Mulholland, eyes twinkling with mischief. "But not in a car."

BY THE SPRING OF 1912, 90 percent of the aqueduct was complete, and Mulholland told reporters that "the end of our task seems fairly in sight." The most difficult and innovative phase of final construction was the installation of the massive inverted siphons—jumbo, airtight pipes by which water would be made to move miraculously up steep canyons and over mountains. Twenty-two in number, these unique hoses were actually welded-together steel pipes, the largest in existence. Ranging in length from 611 to 15,596 feet, and at a diameter of 8 to 12 feet, some were large enough to drive a locomotive through, and their tonnage—some sections weighed as much as 52,000 pounds—required 35 trains of 20 cars each to transport them from steel foundries to the connecting links along the aqueduct line.

They could not be moved by manpower alone, and Mulholland commissioned new equipment to get the job done. The work moved relatively quickly after a steam tractor, dubbed a "caterpillar," was introduced to the world. To move the millions of tons of earth, the steam dredger was another innovation, and the men christened the machine "Big Bill," a joint compliment to engineer Mulholland and the Presidential nominee, William Taft, who himself weighed well over 300 pounds.

Until the machines were later refined, the clanky contraptions broke down often, forcing Mulholland to resort to using old-fashioned but ever-reliable mule teams. Fifty-two mules to a team were needed to haul single pieces of the giant siphons, using three parallel jerk lines of sixteen animals each, with a "lead pair at the head and two wheelers on the tongue." The most famous of the mule skinners, "Whistling Dick," who hauled borax from Death Valley in earlier days, fell from his saddle and was crushed to death in the track of the enormous wagon wheels. He was seventy-four years old.

In March 1913 the Jawbone Siphon, near Water Canyon, the last of the great uphill conduits, over 7,096 feet long, was finished and Mulholland made preliminary preparations to fill the aqueduct with water. He and Harvey Van Norman traveled to the Owens Valley to the headgates of the aqueduct to turn the water from the Owens River into the first intake canals.

Harvey Van Norman, simply known to his friends as "Van," was said by Mulholland to possess "perfect mental equipment for the job of aqueducting." Tall, lean, and handsome, Van Norman projected a genial, honest, and kind personality. Son of a Texas pioneer who had served in the Civil War, Van had become a proficient electrical and civil engineer. He had launched his career as a steam engineer for the Los Angeles Railroad Company and served as superintendent of the Pacific Electric Railway. He had been working as head of construction for the Los Angeles Gas and Electric Company when he was engaged by William Mulholland in 1907 to be electrical construction engineer in charge of building the hydroelectric plant at Cottonwood Creek, where he and his young bride, Bessie, spent the first days of the construction in primitive living conditions. During the next ten momentous years, he and Mulholland would experience the most deep-rooted, poignant friendship and professional association imaginable, and eventually Van Norman would succeed Mulholland as chief engineer for the Los Angeles Department of Water and Power.

At the intake, a small party of celebrants gathered on the concrete floodway. As one of them caught the historic moment on film, Mulholland and Van Norman prepared to open the large iron valve and release the water. Jubilant, Van's wife Bessie smashed a magnum of expensive French champagne against the safety railing. The massive holding gate creaked, and the first gallons of Owens River water surged into the canal.

For the next two days the exuberant party followed the slow progress of the water through the fifty miles of tunnels, canals, and siphons until it flowed into the Haiwee Reservoir. The party dispersed, and Mulholland continued to supervise each step of the water's further advance.

When the flow reached the Elizabeth Tunnel, seventy-five miles south, John Gray and his son, Louis, were on hand to celebrate. The final holding reservoirs that were located at each end of the tunnel were filled with water and the grand opening of the aqueduct was scheduled for July. The water would be stored here until its final journey to the San Fernando Reservoir. As Mulholland and Van Norman congratulated one another and those around them, an urgent phone call came from a frantic engineer up the line at the Sand Canyon siphon. The siphon had developed a leak and was spilling water down the north side of the ravine. Built of two underground tunnels running down and up two mountains, and connected by a steel pipe across a canyon, this siphon was the only one of its kind on the aqueduct.

Rushing back to Sand Canyon, Mulholland and his aqueduct men repaired the leak within forty-eight hours. A small leak occurred on the slope of the south side mountain, prompting Mulholland to test the full pressure capacity of the siphon. He ordered his men to gradually increase the flow of water and waited. When the flow reached 42 second-feet a huge length of the underground siphon was lifted up by the pressure. Water spewed into the air and the canyon wall burst; huge chunks of concrete crashed into the ravine. A workshop was sheared away from the hillside in the avalanche of mud and boulders, and construction equipment was buried beneath debris.

Van Norman was called and arrived the next day to inspect the wreckage. He was harnessed into a chair and lowered by rope into the huge cavity of the ruptured pipe. Above his head as he descended, great chunks of concrete hung from the reinforcing rods. Returning to the surface, Van Norman informed Mulholland that the only choice they had was to rebuild the siphon.

Mulholland agreed, knowing the months it would take to rebuild would delay the grand opening of the aqueduct past summer and would only fuel the opposition forces and give them one more chance to "cry graft" all over again. The collapse of the Sand Canyon siphon was met with disappointment and frustration in Los Angeles as well, and the detailed plans for an elegant summer opening with parades, balls, and gala dinners were postponed indefinitely. Mulholland promised to have the aqueduct ready by Thanksgiving, relieving some of the political pressure, although his engineers remained skeptical. Working with urgency, the crews completed the new siphon by September, and Mulholland announced that the new opening for the Los Angeles Aqueduct would be on November 5, 1913.

The greatest and most controversial water project in North America was finally finished. Mulholland could only grin meekly at his exhausted workers who had gathered around him in celebration.

"That's all, boys. Pick up those hammers and derricks and things and get along home," he said rather sadly. "It's over."

All that remained was for Mulholland to accept the praise, and the people of Los Angeles to reap the profits of his labor and struggle. Completed almost exactly five years after he had first broken ground, and eight years after he and Fred Eaton had first announced the project, the city of Los Angeles would turn to William Mulholland in his greatest moment of triumph and in its own hour of deliverance.

7

DELIVERANCE

I give waters in the wilderness,
and rivers in the desert,
to give drink to my people,
my chosen.

ISA. 43:20

AS THE MORNING SUN began to rise high in the California sky, five thousand wheeled vehicles of every kind—touring cars, roadsters, shiny Big Stephen limousines, underslungs, horse-drawn buggies, hacks, and buckboards—snaked along the already sweltering desert floor, leaving a brown dust cloud twenty miles long.

The massive caravan rumbled north to the natural limestone amphitheater carved in the hillside four miles above the sunbaked hamlet of San Fernando. A dozen steam locomotives brimming with excited spectators squealed to a stop at the little San Fernando railway station and discharged their eager cargo who made their way by foot up to the amphitheater.

By noon, 43,000 wide-eyed men, women, and children, enough people to fill six city blocks, had gathered before the rough wooden grandstand constructed at the base of the amphitheater and garlanded with patriotic red, white, and blue bunting to witness the historic moment. Of this number, 25,000 had come by automobile (the largest number of automobiles ever assembled in one spot in the state of California), 10,000 by train, 2,000 by motorcycle, and 6,000 by carriage, wagon, horseback, or on foot. Clad in their Sunday best—bright bows and knickers, stiff Eton collars and shiny Panama hats, lacy ankle drawers and petticoats—they stood ten deep along the banks of the "Grand Cascade," the so-named graceful, gray concrete trench meandering from the north behind the grandstand and down into the valley.

One out of every five persons living in Los Angeles had come this day, November 5, 1913, to see the precise moment the long-awaited water would at last tumble through the giant sluice gates on its journey into their parched valley and to canonize the man who had delivered them from their thirst.

The official aqueduct program handed out to the sweating throng described in the florid prose style of the time the miracle which they were now about to witness:

> This is the story of a dream that came true; of an idea audaciously conceived and splendidly realized. It is an outline sketch of the history of the Los Angeles Aqueduct, the giant conduit of concrete and steel that brings a river across hundreds of miles of deserts and through mountains to make possible the building in California's sunny southern land of one of the wonder cities of the world. Less then a dozen years ago the vision of the dreamer was told in these words:
>
>> A drop of water, taken up from the ocean by a sunbeam, shall fall as a snowflake upon the mountain top, rest in the frozen silence through the long winter, stir again under the summer sun and seek to find its way back to the sea down the granite steep and fissures. It shall join its fellows in mud follies in mountain gorges, singing the song of falling waters and dancing with the

> fairies in the moonlight. It shall lie upon the bosom of a crystal lake, and forget for a while its quest of the ocean level.
>
> Again, it shall obey the law and resume its journey with murmuring and frettings; and then it shall pass out of the sunlight and the free air and be borne along a weary way in darkness and silence for many days. And at the last drop that fell as a snowflake upon the Sierra's crest and get out to find its home in the sea, shall be taken up from beneath the ground by a thirsty roof and distilled into the perfume of an orange blossom in a garden of the City of the 'Queen of Angels.'

With police sirens screaming, the official motorcade, forty banner-draped black Model T's, crossed over the Southern Pacific tracks and, amid waving American flags and blasting horns, parted the throng and pulled to a halt before the grandstand. Black-suited dignitaries—mayors from San Diego and San Francisco, the governor of California, both senators, various congressmen, and an envoy from Woodrow Wilson—stepped from the cars, waving to the tumultuous crowd. Among them, smiling broadly, were the five men who named themselves the Board of Control. They were an elite group of Los Angeles businessmen who had been accused of manipulating privileged information about Mulholland's "big pipe" to their own advantage. While others in the motorcade had supported or opposed the aqueduct openly, the five men had quietly purchased cheap land options in much of the San Fernando Valley, planning to sell them at hefty profits as soon as this greatest day in Los Angeles history arrived. Dressed in the finest clothes their already considerable wealth could buy, publisher Harrison Gray Otis, his son-in-law Harry Chandler, former water commissioner General Moses Hazeltine Sherman, land developer H. J. Whitley, and Otto F. Brant, vice-president of the hugely successful Title Insurance and Trust Company, made their way to the honored seats on the grandstand to await the arrival of the guest of honor.

William Mulholland was the last to exit his car and climb the stairs to the stage. The throng erupted in a standing ovation, recognizing the fifty-eight-year-old Mulholland immediately from the many photographs of him they had seen as they excitedly followed the progress of his aqueduct in the newspapers. He was everything they imagined their hero to be—six feet tall with a bushy black-and-silver mustache, he projected the charm, energy, and vigor that earned him the affection of all those who had worked under him and now the awe of a public as well.

William Mulholland was a man whose truest feelings, even ordinary daily reactions, were hard to gauge and rarely expressed. Although an aficionado of the arts, he was first of all a doer, a man of action, and action guided his life. Like most men born in the mid-nineteenth century, to speak of his inner self was foreign to him—inconsequential things like fears or yearnings were a waste of time and best left unsaid. It was one thing to be frank and blunt to get the job done, but little in a man's life could be improved by expressing self-doubt or weakness. The tough workaday world required single-mindedness of purpose. Nearly all of his professional life was devoted to the relentless pursuit of water. His sole interest was in advancing the public good and fulfilling his vision—to make desert-locked Los Angeles into a thriving metropolis.

Even on this glorious day, Mulholland's face was controlled and measured, and amid the continuing ovation, he walked straight and tall to his honored position on the grandstand, every inch a hero. Never before had a city of 250,000 put so much faith in one man. This was the capstone of his life, a moment of triumph as few have known. Yet, beneath the calm, sun-hardened features was despair. Back in Los Angeles, Lillie lay dying, and Mulholland knew that neither the cheering of the crowd nor his resolve could save her.

One by one, the dark-suited officials came forward to the rostrum and heaped unstinted, eloquent praise on their calm, efficient engineer. "William Mulholland . . . has given of himself and of his best in unswerving loyalty to build the aqueduct for the people of Los Angeles. Through the long years of toil and planning, fighting against obstacles, boring through mountains, and bridging deep canyons, Mulholland continued his work without excitement or flurry," droned but one of his many admirers to the restless crowd.

At long last, speaker William Kinney of the Chamber of Commerce came forward to introduce the hero of the hour:

> We came to believe that not only could the city have the water but that we could get it by spending money and building a great conduit. Then the people began to ask where a man could be found who was big enough to tackle the job and put it through. We found him. He was right there in our midst. We decided that Bill Mulholland was that man. And we never changed our mind. I have worked in the some of the big cities of the world. In Washington and elsewhere I have come in contact and have known men in high station and engineers of prominence, and I have never met a nobler man, or a kinder man, nor a better engineer than William Mulholland. And here he is!

As Mulholland rose from his chair and moved to the rostrum to speak, the crowd surged forward, cheering, waving a sea of tiny American flags in time to the salutes of National Guard rifles and the crashing of military band brass cymbals. Dozens of newspaper cameras flashed as one, billowing spent flash powder into the air.

"Ladies and gentlemen," Mulholland began, his strong voice booming across the now-hushed crowd, "in the few remarks I shall make, I speak not for myself alone but for my associates of this great work."

With his voice growing hoarse in the dusty parched air, he went on to praise the army of men who labored beside him at the drills in the tunnels, at the cement molds, the men who dug the ditches with great steam shovels, the muleskinners, the riveters, the forge men, the blacksmiths, masons, skippers, swampers, blasters, and engineers:

> They do so much for so little. I know this type of man from my early life as a sailor and worked with them, slept with them and I would rather sit around camp with them than be in a circle of lawyers, doctors or bankers. Professional men are trained to conceal their thoughts but these men are frank, blunt and human and a

man gets more real insight into human life and affairs with them than with the other type. They were a grand lot, they did their work and took their chances in the tunnels, dry or wet, safe or indifferent, with gas or free from it and in other dangerous jobs and they spent their money like sailors ashore and that is the one thing that saddens me today. It has been a close partnership and we have worked together well. Therefore we appear jointly and this expression is on behalf of all of us.

Giving final tribute to his loyal aqueduct workers, Mulholland paused to clear his throat. Magnanimously he praised the forgotten man whose idea the aqueduct had been. Calling Fred Eaton "the father of the big ditch," he honored him simply and honestly, making no mention of the private differences that now existed between the two men. "To former mayor Fred Eaton, I desire to accord the honor of conceiving the plan of the Aqueduct and of fostering it when it most needed assistance," he said in closing.

Notably absent from the ceremonies—in the official aqueduct program, tribute was made to Eaton's conception of the aqueduct and his photograph appeared next to Mulholland's—Eaton had offered the excuse that heavy autumn rains had so badly diluted the aqueduct water as to make the first drops "untrue." Since the aqueduct was largely carrying rainwater, it would be a farce to attend the celebration, he claimed. After his three visits with Mulholland proved fruitless, and he realized that, at least for now, his price for Long Valley would be refused, Eaton's refusal to attend the ceremony was another attempt to antagonize his former friend. Exacerbating the ill feelings between the two men, and knowing full well that his remarks to the press would get back to Mulholland, Eaton's boycott of the ceremony was a childish jab at diminishing Mulholland's pleasure in the celebration.

At a huge roar from the crowd, Kinney stepped forward and presented Mulholland with a handsome, engraved, silver loving cup. "Ladies and gentlemen," Mulholland said, continuing with his praise, and holding the cup up high for the cheering crowd to see,

This is yours. It was your own fidelity and unfaltering courage that made the work possible, and I want to thank you. This period in my life is one of great exaltation. The aqueduct is completed and is good. No one knows better than I how much we needed the water. We have the fertile lands and the climate. Only water was needed to make this region a rich and productive empire, and now we have it. This rude platform is an altar, and on it we are consecrating this water supply and dedicating this aqueduct to you, your children and your children's children—for all time.

And at this he lifted his hand in signal, and the crowd, holding their collective breath, waited, their eyes now glued to the giant cast iron wheel that would lift the gates, allowing the water to enter the bone-dry canal.

Stepping to the flagpole mounted on the platform, Mulholland pulled the lanyard and unfurled the stars and stripes. Los Angeles's own coloratura diva Ellen Beach Yaw's clear soprano voice rose above the noise of the crowd and the band blared "God Bless America," fireworks thundered, drums rolled, dusty Panamas, dainty white handkerchiefs, and programs were thrown into the air and fell like snowflakes upon the grandstand. The greatest moment in Los Angeles history had arrived.

Manning the wheel, General Adna Chaffee and Van Norman struggled to open the enormous spigot. On command of an officer, the military men began firing field guns in salute. The crowd packed itself in even tighter alongside the sluice gates—risking their lives on the narrow, three-foot-wide concrete aqueduct embankment with its sixty-foot sheer drop. Many clung to the long wire fence along the embankment to avoid falling. Captivated by the suspenseful moment, none took their eyes off the mammoth gates, and so failed to see the impervious William Mulholland, the greatest water engineer of his age, standing there, one hand held behind his back with fingers crossed in hopeful good luck. He then lifted his eyes heavenward in thanks as an awesome creaking noise was heard and the gates slowly rose and the water, sparkling like diamonds in the autumn sun, squeezed its way through the concrete funnels into the slide.

At first the crowd saw only a trickle, which suddenly became a stream and then a raging torrent as it flowed in the culvert below them. From high in the snow pack of the Sierra Nevada 233 miles north all the way to the San Fernando Valley, eight thousand miner's inches were pouring from the hatches and splashing down the chutes in a veil of spray above newly man-made falls. Forty-three thousand hearts were beating a little faster as the daringly conceived and boldly executed project—one of the great engineering feats of all time—was successfully completed. And almost in a flash of an eye, there was delivered to the people of Los Angeles an asset worth a hundred million dollars—four times the cost of the aqueduct. It brought assurance of metropolitan grandeur and future prosperity such as few cities of the world can hope to attain.

The crowd went wild at the sight of the water. Mulholland turned to them and over the roar of the rushing water, shouted, "There it is—Take it!"

And they took it. Excited children raced down the concrete incline at its lowest point, following the first foaming water through the gates. They were joined by happy thousands, equally excited, who waded in the cascading Owens River water, splashing and jumping in exuberance over their deliverance. Only water was needed to make Los Angeles a rich and productive empire and now, as Mulholland announced, here it was—and the people, as predicted, came to take it. In just seven years the population doubled to exceed half a million, and it doubled again in the next decade. Today that population has increased to three and a half million people.

As the military fired a twenty-one-gun salute and the band struck up "Hail to the Chief," Harvey Van Norman pushed his way through the throng and up the steps of the grandstand to Mulholland, who was watching the joyous scene in silence. Above the din, Van Norman whispered something in Mulholland's ear that caused him to break out in a smile from ear to ear. Van Norman had told him that the doctors in Los Angeles had wired to say that Lillie was faring well and resting comfortably.

In this moment, with the full realization that his battle had been fought and won, with the crowd cheering and exulting, emotion overcame him. High-ranking officials, many of them among his closest friends, clapped him on the back and congratulated him, but for a moment he could not answer, remembering the long years of struggle and work. When he finally did speak, it was only to say briefly and modestly, "We knew it could be done and here it is."

The reputation of Los Angeles as a colony of Hollywood, doomed to insignificance by its dependence on water, was changed forever. The city could now deliver what it promised, a Mediterranean paradise sprung from a lifeless desert.

"This is a great event," Mulholland declared, "fraught with the greatest importance to the future prosperity of this city. I am overwhelmed and honored. What greater honor can any man ask than to have the confidence of his neighbors?"

Suddenly, Mulholland's measured and sober expression gave way to a slight smile. Tears streamed down his face. Then the gruff, hearty immigrant laughed out loud, and he marched down to the water's edge to join the crowd frolicking in the Owens River water.

THE FAMILY THAT DAY was represented by two of Mulholland's daughters. A week before the big event, Lucille, age seventeen, and Rose, twenty-two, accompanied by Bessie Van Norman, visited an exclusive department store on Wilshire Boulevard in downtown Los Angeles, and purchased new dresses. Lucille selected a white, short-sleeved, high-waisted cotton frock, and Rose picked a more austere, jacketed daytime suit. Bessie Van Norman, recognizing the occasion as one of the peak moments in her husband's life, splurged a little, and chose a stunning, Egyptian-cotton opera wrap with lace spats and ankle-length drawers, worn with a beige silk hat trimmed in liberty green, gloves, and matching chiffon scarf.

As taught by their father, the Mulholland girls were frugal and conservative, but they knew they would be photographed, and like Bessie spent more than they had planned, wanting to look their very best.

Lillie was too ill to attend the ceremony, and remained under her doctor's care in her blind-drawn bedroom, while Lucille and Rose were driven to the opening in a department car, traveling separately from their father, who had left at daybreak in another chauffeured limousine with other dignitaries. Fifteen-year-old Ruth, the youngest daughter, and Mulholland's pet, remained at home at her own insistence to tend to her mother.

The day before, Lillie had suffered respiratory failure and had been rushed to the Queen of Angels Hospital, where doctors successfully resuscitated her. Desperately wanting to be with her husband in his finest hour, Lillie insisted that she be sent home. She had planned to accompany Lucille and Rose up to the aqueduct, and they had helped her bathe and dress, but ultimately she proved too weak to make the arduous trip. Unaware of her intent, Mulholland had left strict orders at the house that word of any change in her condition be sent to him posthaste at the ceremony.

Now sitting at the head of her mother's bed, slowly brushing her long white hair, Ruth joined her mother picturing the events of the great day. Lillie read from the aqueduct program, knowing that at this moment the water was already on its way to Los Angeles.

When the Mulholland daughters returned home that night, their arms were filled with souvenirs—pennants, brochures, gold-trimmed ribbons, tiny vials of Owens River water, and a Panama hat. They all sat on Lillie's bed and watched Rose place the small items into their father's scrapbook, one of the treasured possessions of the Mulholland family.

On Thursday, November 6, the morning following the celebration, the Los Angeles newspapers were plastered with pictures of the grand opening and of the two Mulholland daughters. One photograph of Lucille, arm outstretched clasping a silver loving cup filled with Owens River aqueduct water, was printed nine inches tall in the *Los Angeles Herald*. In the *Los Angeles Times*, the two sisters were photographed hugging one another, smiling under their bonnets, teary-eyed and waving white handkerchiefs, as the first water crashed into the conduit.

Also not in attendance was Mulholland's oldest son Perry, age twenty. Working with a group from the U.S. Geological Survey that was conducting its first assessment of the White Mountains north of Panamints in Inyo County on the day of the grand opening, Perry wrote his family how pleased he was that the aqueduct was finally finished. "Nothing will please me more than to hear that the knockers have had the quietus placed on their yapping and howling for all time to come, as far as the Aqueduct is concerned. I'll bet Pappa isn't shedding any tears tonight after having such a heavy load lifted from his shoulders."

Young Mulholland was right. For now, the public yapping and howling for Mulholland's head had disappeared in the flow of the aqueduct water. As forty-three thousand sunburned, tired but jubilant onlookers began their long trek back to Los Angeles, Mulholland in his limousine, followed by two hundred select guests, proceeded to the spacious ranch home of Fred Bouroff, one of the major boosters of the San Fernando Valley, where a reception was given in Mulholland's honor.

The elaborate reception, financed by $500 from Bouroff and $300 donated by the Board of Control, was only a small taste of the homage that would be paid to Mulholland in the coming days. Hardly had the guests assembled when they gathered around him and joined in a rousing chorus of "For He's a Jolly Good Fellow." Mulholland was presented by Bouroff with yet another set of handsome silver loving cups, each bearing an artistic engraving of a construction phase of the aqueduct.

"Your deeds and example will last forever," said Bouroff, handing the cups to Mulholland.

This was followed by much applause from the assembled supporters and by the serving of much champagne. Although joyous, it had been a long, hot day beneath the desert sun, and Mulholland, who was known to leave a trail of "dead soldiers" when he was in a thirsty mode, drank heartily and often.

The praise continued non-stop from the various dignitaries. In keeping with his character, Mulholland received the accolades with modesty.

"The magnitude of this generous reception afforded me today is embarrassing," he said in his formal, slightly askew habit of speaking. "I have done no more than my duty, and with the confidence and trust reposed in me by the public, no man with a soul in him could do less than respect it and make good."

At this, another round of applause ensued, and even more drinks followed. Mulholland continued, "I feel that the praise you are giving me is in large measure due to those who have so magnificently assisted me in this work. I cannot forget the unfaltering championship of former mayor Alexander and General Chaffee. He has served his country with credit in large affairs, he gave of his time and mind in assisting me in this great work."

Cries rang out for General Chaffee to speak, and he obliged. "There is nothing to equal this in magnitude this side of the city of New York. Notwithstanding the disclaimer of Mr. Mulholland I insist that to him is due the credit of this achievement. We could only do our duty by upholding his hands. It was his genius that conceived and his skill that executed this great work. Ages ago Moses smote the rock and pure water rushed forth. Seven years ago we smote the rock and $24,000,000 came forth and today the poured water flows forth which will enrich and make happy the half million persons here and the half million who are soon to come."

Thunderous applause from the boosters present shook the house, and Mulholland, always equal to the situation, stepped forward again on tipsy legs and made his final pronouncement.

"Failure cannot come to anything that southern California undertakes with such citizens. We are undoubtedly a people doomed to success." And in tribute to Mulholland's great feat and his unusual toast, the crowd shouted in laughter.

The city of Los Angeles celebrated the completion of the aqueduct with religious zeal. A series of official dinners, balls, and parades coincided with the events at the Cascades, and an expensive plaster model of the aqueduct was built at Exposition Park so that citizens could view in miniature the project they had so courageously funded. Souvenir vendors did big business dispensing American flags and thousands of small bottles of the first aqueduct water to flow through the gates. A schedule of gala public events for the day following the opening was published in the *Times* that included a series of elaborately produced band concerts at Exposition Park's Sunken Garden, a roaring gala reception, dedication of an aqueduct memorial fountain by a U.S. senator, the laying of a commemorative cornerstone at the State Seventh Regiment Armory, followed by another address officially opening the Los Angeles County Museum of History, Science, and Art, athletic games, and ending with a reception for distinguished guests hosted by the Board of Public Service Commissioners.

All through the week the celebration continued, as did the praise for William Mulholland. The waterway was immediately recognized worldwide as the single greatest water project of its time. And the Chief, showered with honors and awards, was proclaimed everywhere as the West's greatest man. Woodrow Wilson, Will Rogers, and other famous men heaped unstinting and well-deserved praise upon him. The *Los Angeles Times* paid its homage: "William Mulholland, the master of the aqueduct, the peer in practical results of the world's best engineers: every man, woman, and child acknowledges a debt impossible to pay."

A hymn devoted to the aqueduct was published in the *Evening Herald*:

For I carry a tale of promise to the cities and haunts of men;
Painting the picture of acres—fulfilling my mission
and then—
Straight from the heart of the mountains, where peaks kiss the blue
of the sky;
Born in the mists of the morning—to the sea I have
come—and to die.

From the *Daily News*:

Los Angeles sent forth her engineers; their instructions were to find water, plenty of water, the best of water, and complete their

> plans for bringing it any distance that might be necessary, to serve in abundance the needs of this city for generations. That was eight years ago. Today Los Angeles is celebrating the arrival of the water. There is no longer the trace of a shadow on the destiny of this wonderful city—all due to our Chief William Mulholland.

People by the hundreds came up to him and clapped him on the back and congratulated him. Through it all, Mulholland remained calm and reserved, always giving credit to others.

"The work now stands complete," Mulholland wrote. "The thousands of men who have labored under adverse conditions of desert heat in the eight years since the work was started have laid down their tools. Their reward has been that of a work well done and the confidence of a people not misplaced. They have no excuses to offer; no boasts to make. Their work will live after them, which is their reward."

It was a glorious time for the once-impoverished immigrant from Belfast. He had reached heights ordinary men could only imagine. As reporters clustered around him, he was asked would he ever return to his native Ireland now that the aqueduct was finished. "I never want to see the damn island again," he said.

"How about mayor? Will you run for mayor?"

"I'd rather give birth to a porcupine backwards than be mayor of Los Angeles," he said fervently, exhaling a cloud of cigar smoke.

The idea of running for political office was far from Mulholland's mind, and like many of the rough-hewn workers along the aqueduct line, he had a distrust of politicians and never imagined himself in that role. Still, a serious public effort was launched by political progressives to initiate his candidacy for mayor. Harrison Gray Otis and publisher E. T. Earl openly urged Mulholland to seek the office. Mulholland, flattered, gave his usual cynical response to the proposal. He despised politics and disdained politicians even more.

But as the waters roared into the San Fernando Valley, charges that Mulholland had conspired with the Board of Control to deliver water to the board's arid valley lands resurfaced. Influential critics continued to allege that Mulholland had helped conceive the aqueduct to benefit the powerful members of the San Fernando Mission Land Company. Once the water had been successfully delivered into the northeast corner of the San Fernando Valley, Mulholland never escaped the stinging accusation that the aqueduct was built solely for the greedy exploitation of the Otis-Sherman-Huntington land syndicate.

Mulholland's work on the aqueduct was well under way when a second syndicate was formed in 1909, this time comprised of Harrison Gray Otis, Harry Chandler, Otto F. Brant, Hobart J. Whitley, and Moses Hazeltine Sherman. This new syndicate, called the Los Angeles Suburban Homes Company, or simply the Board of Control, secured an option on 47,500 acres in the San Fernando Valley at a total price of $2.5 million. Together, the two syndicates' tracts encompassed nearly the entire San Fernando Valley from the present site of Burbank on the east to Tarzana on the west, including what is today known as Van Nuys, Canoga Park, Reseda, Sherman Oaks, and Woodland Hills. The company exercised its option in 1911, two years before the completion of the aqueduct, filing a subdivision map for "Tract 1000," the largest single land development in Los Angeles history. Immediately the Pacific Electric Railway began construction of an extension into the huge new subdivision, virtually assuring its financial success.

The men in the Board of Control were the power oligarchy of southern California and world-famous, each having amassed a private fortune. Harrison Gray Otis, who owned the *Los Angeles Times,* had come to Los Angeles in 1881. Known for his tightfisted, stingy personality, Otis loathed men disdainful of growth. "Hustlers, men of brain, brawn, and guts" were the people he admired. A "large, blubbery man with a Bismarck moustache," Otis was never a subtle negotiator and was known for his violent temper, later earning him a reputation as Los Angeles's most disliked capitalist. The greatest experience in his life was as a Union soldier in the Civil War that "transformed him from farm boy turned printer into a man who, once having experienced power, lusted after it even more." Dubbed the "walrus of Moron-Land," Otis had come to California broke and within one decade established himself as one of the city's most influential and overbearing citizens.

"There are certain combinations," wrote author William G. Bonelli, "either of men or things—which appear to have been ordained by some natural law. These would include ham and eggs, hot dogs and mustard, Sodom and Gomorrah, and Harrison Gray Otis and Harry Chandler." Harry Chandler, born in New Hampshire, dropped out of Dartmouth College after he jumped into a starch vat on a dare and developed a severe lung infection. He migrated to Los Angeles for its climate and found his first job picking fruit in the San Fernando Valley. Chandler held onto his money, later purchasing newspaper circulation routes and then independent newspapers. In 1866 he purchased Otis's rival paper, the *Tribune*. To deal with Chandler, Otis purchased Chandler's distribution system, shut it down, and hired him as senior editor. Within two years the ubiquitous Chandler became Otis's son-in-law.

Moses Hazeltine Sherman had been a general in the army at twenty-nine, fighting Apaches in the Indian Wars. He moved to Los Angeles from Arizona, where he had been a school administrator, and later built the first rail line in the city—the famed "Red Car—" becoming a wealthy and eccentric trolley magnate. Otto F. Brant was vice-president and general manager of the highly successful Title Insurance and Trust Company. Originally from Ohio, Brant came to Los Angeles in 1888, becoming a shrewd, visionary real-estate speculator who was credited with the creation of the escrow process.

The fifth member of the board, H. J. Whitley, was known as the "Great Developer." Born in Toronto in 1860, Whitley was the youngest of seven children. At eighteen, he left Canada and pursued a career developing towns across the West. He constructed small cities from the Great Plains states deep into South Texas, making him a legend among developers and financiers. He was a friend of Teddy Roosevelt and an original Sooner. His most famous development, an obscure ranch spanning the hills just north of Los Angeles, was an orange grove he bought in the 1880s, later named Hollywood. Whitley believed that "land without population is a wilderness, population without land is a mob."

Pro-conspiracy advocates described Whitley as "the chairman," Sherman as "the spy," and Otis as "the booster" who chartered the board's course. Together, it was alleged, they swayed public opinion, influenced city government, and manipulated information for their own purposes. Mulholland's detractors claimed that Mulholland had steered the aqueduct to end in the San Fernando Valley, thereby guaranteeing each man in the syndicate a massive fortune after the water arrived. After the opening at the Grand Cascades, land values boomed. Within two decades, land values jumped from $20 to $2,000 an acre, putting estimated syndicate earnings at $100 million.

The Board of Control, like the earlier syndicate formed in 1905 (the San Fernando Mission Land Company) represented a microcosm of the larger Los Angeles business community, and members of both groups overlapped. According to historian William Kahrl, the divisions within the group were as revealing as its alliances:

> Huntington despised Sherman personally, for example, and would be betrayed by him in a business deal only a year later. Otis had fought Huntington bitterly over the issue of the city's harbor at San Pedro. Otis also regarded E. T. Earl as his most deadly competitor in the local newspaper industry, and the two devoted ten years of their lives from 1901 to 1911 trying to drive one another out of business. And Huntington and Harriman, of course, had been at war since 1900, if not before.

As heated as the rivalries were, members of both syndicates united on issues where their best interests were at stake. Support of Mulholland's aqueduct was a primary example.

Charges that Mulholland was engaged in San Fernando Valley land "thievery" never subsided. Yet, despite attempts to link him with the wealthy financiers of the syndicates, as one historian has noted, "he never really belonged in their world, in their clubs, or their way of life . . . and seems in general to have been rather careless about money."

Mulholland's philosophical inclinations leaned more toward public service and less toward personal gain. "A man's worth is measured by his importance to society and to humanity generally," he insisted. "I never wanted to be wealthy. All I did want was work." But such civic aspirations seemed disingenuous to Mulholland's enemies. To the end of his life, Mulholland continued to insist that he was never involved in land speculation, and had no ulterior motive in the aqueduct's construction.

To his credit, Mulholland provided solid reasons for the proposed aqueduct to end at the northern end of the San Fernando Valley rather than proceed to the Los Angeles city limits. One compelling consideration was that the San Fernando Valley constituted a vast underground reservoir which could store Owens River water and raise the water table. The valley was the best receiving basin—water deposited there would automatically drain into the Los Angeles River and its broad aquifer, creating a vast, non-evaporative storage pool for the city to tap as needed. Further, it was a location where power plant construction was entirely feasible.

Even though he tried to distance himself from the Board of Control, charges that he engaged in land speculation never ceased to haunt him. Historians later documented that no grand conspiracy between Mulholland and the land syndicate existed. Only a circumstantial case against Mulholland could be made, and no hard evidence ever surfaced proving that either syndicate did anything other than what they were in the habit of doing—investing purely for speculation. But until the day he died, Mulholland was found guilty by association, accused of land thievery and profiteering in connection with the Los Angeles Aqueduct.

PRODIGAL DAUGHTER

A man shall be known
by his children.
ECCLES. 11:28

IN APRIL 1915 Lillie Mulholland lapsed into a coma at the German Hospital and never regained consciousness. On April 28, she died at age forty-seven from cancer of the cervix. Mulholland was with his wife during her last excruciating days, and despite its inevitability, her death hit him especially hard.

Lillie had always been reluctant to participate in her husband's public life, and during their marriage she maintained a retiring disposition, never caring to share in the public spotlight. Like the wives of many great men, her identity was obscured by the long shadow of her husband's success; the *Los Angeles Times* summarily described her as the "devoted loyal wife of one of the foremost residents of Los Angeles, and the most widely known engineer of America, Chief Engineer, William Mulholland." When the newspaper wanted to print an obituary photograph there was none to be found. Extremely shy and self-effacing, Lillie chose obscurity for herself and refused to pose for photographers, and only tolerated herself to be photographed in the Mulholland family portrait.

In death she received more attention than she would have preferred. Massive flower arrangements were sent from the Chamber of Commerce, the Board of Public Works, the Water Department, and from Mulholland's many friends. The lavish displays paid tribute to Lillie and homage to her husband, the city's powerful and leading citizen. Bessie Van Norman sent a large bouquet of wildflowers hand-picked from the ranch in Chatsworth, mixed with a store-bought bouquet of red roses and white gladiolus. Typically, the modest Lillie had requested that music not be played at her funeral, and the simple Episcopal ritual was followed by attendance at the Rosedale Crematorium. The final service was private to the family. Burt Heinley, Mulholland's secretary, and H. A. Van Norman served as pallbearers as William Mulholland laid his wife of twenty-five years to rest.

Although Lillie had lived to see the beginning of her husband's climb to considerable national prominence, she did not live to see him achieve the status of an international celebrity, nor share the tragic climax of an amazing career.

MORE THAN ANY OTHER ENGINEER of his era, Mulholland revolutionized construction and engineering practices throughout the Western hemisphere and much of the world. He was the first engineer in America to make practical use of hydraulic sluicing, a technique used in the construction of Silver Lake Reservoir. The concept was revolutionary and attracted the attention of engineers nationwide. In 1906, he instructed government engineers in how to move material over long distances through a modified method of hydraulic sluicing, and his method was adopted in two of the Panama Canal's three most difficult construction problems—Gatun

Dam and Culebra Cut. Since then, the techniques have been used extensively in cities around the globe.

Mulholland was also the first American engineer to make major use of hydroelectric power in construction, the first to revive the old Roman method of making cement (saving three quarters of a million dollars in costs to build the Los Angeles Aqueduct), the first engineer in the world to use caterpillar tractors, and was a pioneer in tunneling machinery and methods, and an innovative leader in design of high earth dams. By 1913, he had successfully designed and constructed "the most gigantic and difficult engineering project undertaken by any American city" as well.

Mulholland's achievements brought the Irish immigrant academic honors he never dreamed possible. The University of California at Los Angeles awarded him an honorary Doctorate of Laws in 1915 in an elaborate ceremony hosted by the Board of Regents. Honorary memberships were bestowed on him by the American Water Works Association, the National Association of Power Engineers, and the Tau Beta Pi engineering fraternity. For the next decade, the name William Mulholland was emblazoned in headlines and he was photographed, quoted, sought after, and showered with civic awards wherever he traveled.

Mulholland's huge success in conquering the manifold problems of the aqueduct endeared him to the polyglot people of Los Angeles. Like a child who needed a baseball star to idolize, the city needed a mythic figure, and the newly crowned hero represented to Los Angeles citizens an "awe-inspiring glimpse of perfection." To them, as one devotee wrote, Mulholland embodied a "constellation of super-human traits and unlike many of today's celebrities, his achievements were built through hard work, and were genuine even in the scrutinizing eyes of God."

Author Joseph Campbell wrote that the hero is usually the founder of something—the founder of a new age, the founder of a new religion, the founder of a new city. William Mulholland was the realization of the American ethic of industry and self-education, an original American persona who became an integral part of the mythology surrounding the creation of modern Los Angeles. Rugged, fearless, and determined, he was a self-realized man of the West, a genius of the people. Like many of them, he had been a mere immigrant starting out with nothing more than a strong back and a willingness to work.

As a celebrity, Mulholland was a superstar in that era's version of the national lecture circuit, speaking to groups of every kind that solicited his appearances and paid handsome lecture fees and travel expenses. He was an exceptional public speaker, and the many thousands who heard him could testify to his uncanny ability to move an audience to laughter and tears with his mesmerizing voice and odd, formal language. His great store of anecdotes and his sassy wit made Mulholland a genuine crowd-pleaser—the kind of speaker that was always saved for last in an evening's program. Mulholland's earthy speeches and popular public appearances played a significant role in creating his public mystique, something that today's image-makers could envy.

As his fame and prominence grew, journalists found Mulholland to be a difficult interview. His charismatic personality was often overbearing, and they complained that Mulholland dominated every discussion, never seeming to answer a question directly. More than one exasperated journalist said that Mulholland's thoughts jumped from "psychology to international relations, from etymology to Sarah Bernhardt, from Rembrandt to baseball," all in one sentence. Nevertheless, requests for interviews continued to pour in, and Mulholland's secretary, Burt Heinley, assisted him in answering the enormous amount of correspondence, scheduling interviews and assisting in writing many of Mulholland's speeches.

As one reporter noted, a big part of Mulholland's appeal was due to the fact that even though he was an engineer, he had none of the characteristics of the typically boring technician. He was, by all accounts, larger than life, and he credited his engaging personality to his theory of "mental expansion." Mulholland explained to one journalist that the reason engineers were considered dull was because they made no effort to "broaden mentally in any other direction but by their slide rules."

"The only feasible way to study mankind is reading good books, written by men who were masters of their art," he pontificated.

Mulholland's reading preferences included authors like Shakespeare, Twain, and Carlyle. "No one," Mulholland snorted, "can say that he understands human nature, the actions and reactions of men, unless he had read Carlyle's *History of the French Revolution.*"

"Damn a man who doesn't read books. The test of a man is his knowledge of humanity, of the politics of human life, his comprehension of the things that move men," he once exclaimed to yet another weary reporter.

Mulholland's public image was cultivated not only in the many colorful articles written about him, but also in pieces that Mulholland wrote himself. In addition to the dozens of articles about the region's water supply, the construction of the aqueduct, and his vision for a second aqueduct which were published in newspapers and trade journals including *Hydraulic Engineering, Journal of the American Water Works Association, Community Builder,* and *Municipal and County Engineering,* lively articles for the non-engineer promoting various controversial Mulholland projects were often published with considerable attention in Harrison Gray Otis's (later Harry Chandler's) *Los Angeles Times.*

Despite the saintly reverence bestowed on him, people knew Mulholland was very much a human being. His vices were known to be excessive drinking, smoking, and a reputation for telling obscene jokes when women were not present. He would occasionally lose his temper and verbally abuse his children and closest associates. He had little patience with others—including young children. Once, a group of elementary school children on a field trip observed a flock of geese swimming at the Haiwee Reservoir. Prompted by their teacher, the children badgered Mulholland over what could be done to keep the birds from polluting the water. "Well goddamn it," growled the short-tempered Mulholland, "I suppose we could catch them all and put diapers on them."

For a long while after Lillie's death, the grieving Mulholland was often depressed and out of sorts. Yet despite his crankiness, Mulholland's work habits remained exemplary. He arrived promptly at the office each workday at 7:00 A.M. Today he would be branded a workaholic—in fifty years he never took a vacation—but to him, work was pure pleasure. "Don't forget," he told one young engineer, "I have been working all this time on the work that I loved. That is one of the greatest things that can come to any man—to do all his life the work he loves." Mulholland did, however, break away from his many inspection tours of the aqueduct to go fishing with Dr. Taylor, and on Saturdays during the summer months when he was in Los Angeles he always headed for Hollenbeck Park to play baseball.

After Lillie's death, Rose assumed her mother's role. She took over the many household duties, lovingly tending to her father's every need. Like her mother, she was devout, quiet, and plain. Rose's life, like that of her mother's, was obscured by her father's giant shadow.

Hoping that a change of scenery might help her father's depression after Lillie's death, Rose suggested that Mulholland sell their Boyle Heights home, where they had lived for over twenty years, and choose another elsewhere. Reluctantly, Mulholland agreed to do so, and when Rose found an appealing listing on St. Andrew's Place in the Sunday paper, she and Mulholland drove to see it. When they got there, they found it vacant and the door ajar.

"Let's go in," Mulholland said adventurously.

"Oh no we can't do that, it wouldn't be right," Rose cried, horrified at the idea of entering a house without permission. Over her protestations, Mulholland took her by the hand and went inside.

As they were looking around upstairs, the owner entered and, not recognizing Mulholland, demanded to know who they were and what they were doing there.

"What else? We're looking it over to buy," Mulholland snapped, "and we like it except for one thing."

"What is that?" the man asked, now interested. "Whatever it is it can be fixed."

"No, it can't be fixed," Mulholland retorted firmly. "It has to be changed completely. I want a straight staircase, not a circular one."

The owner looked at Mulholland as if he were crazy. "It's a beautiful staircase. Why on earth would you want to change it?" the man asked perplexed.

"It's obvious. I'm going to live here until I die, and they're going to have a hard time getting my big body down that narrow staircase," reasoned Mulholland, forever the engineer, having made the calculations mentally when he had walked up the stairs with Rose.

Agreeing to the change, the owner asked Mulholland what kind of terms he wanted.

"Terms? What do you mean by terms?" Mulholland growled. "If I can't pay cash, I won't buy the damn thing."

Mulholland left a check for fifteen hundred dollars, paid the balance the following morning, and soon moved into the house. But Rose's high expectations that the bright, sun-filled Victorian house on tree-lined St. Andrew's Place would bring happiness to her morose and grieving father proved wrong; while he resided in the St. Andrew's home, Mulholland faced the most difficult and arduous years of his life.

ON JUNE 15, 1915, Mulholland's daughter Lucille accidentally poured oil onto a hot stove and set the home of Mulholland's sister-in-law, Mrs. L. F. Mitchell, on fire. The blaze spread so rapidly that nothing in the house could be saved. Lucille and her aunt escaped without injury but a trunk of Mulholland's treasured family keepsakes were destroyed. Mulholland accused Lucille of deliberate carelessness and his anger was reported in the newspapers.

It was the first public event in a see-saw of interaction between Mulholland and his impetuous, attractive, modern-minded daughter. The destroyed house was covered by insurance, but the sentimental keepsakes, including those of his deceased brother, Hugh, could never be replaced, and it took months before Mulholland could forgive his daughter for her "stupidity."

Lucille, Mulholland's second daughter and fourth child, had already developed a reputation as the irresponsible sister. Rose, the firstborn, was Lucille's opposite. Her world centered around the approval of her mother and father, and she spent her life in service to them. Rose exhibited a deep-seated dependency on her father which Lucille managed to escape. Even as an adult, Rose was constantly driven to seek her father's approval and love. Lucille, on the other hand, did pretty much what she pleased, which generated storms of conflict not only with Mulholland but with Rose as well.

Lucille's sense of self-esteem was not rooted in her father's opinion of her actions; she didn't place much value on her family's approval, and entangled herself in romantic liaisons that wreaked havoc in the Mulholland household. Mulholland's successful efforts over the years kept his own private life out of public view, but the actions of the impetuous Lucille put gossip of the Mulholland family on the front pages of Los Angeles newspapers, causing the now-image-conscious Mulholland much stress and embarrassment. Mulholland was a classical patriarch: He had raised his children and demanded their strict conformity and obedience. But the intelligent and mischievous Lucille challenged every conventional notion that Mulholland had attempted to instill in his children about discipline and morality.

Lucille was a rebellious daredevil, and more socially adept than her brothers and sisters. These traits, coupled with her pretty face and the Mulholland name, caused many men to seek her attentions. Unlike the other children, Lucille was likely to transgress social conventions and break with conformity. She was the family's free spirit, and had little compunction in divulging family matters to her friends and acquaintances which Mulholland would have preferred to keep hidden.

Older sister Rose and brother Perry were so anxious to please their father that, in deference to his wishes, they made decisions which would later prove to be detrimental to their own futures. Rose devoted her life to caring for her father and died, at age eighty-six, a childless spinster in her father's home. Perry gave up a long-wished-for university education to manage the Mulholland ranch in the northwest San Fernando Valley.

Lucille, in sharp contrast to her siblings, entered into a series of stormy love affairs. A year after her mother's death, she eloped with Edmund G. Sloan, a roustabout at Standard Oil Company, to the shock of everyone in the Mulholland household. By October 12, 1916, one month and four days following her secret marriage, Lucille, aged twenty, had separated from her husband and returned to her father's home, pregnant.

In June 1917, Lucille gave birth to her only child, Lillian E. Sloan, and sought her father's assistance in obtaining a divorce. Sloan had refused to provide for his wife and child since the couple's separation, and, charged with "idleness, profligacy, and dissipation," failed to appear in court.

According to front-page stories in Los Angeles dailies, the couple's split occurred when Sloan demanded that Lucille buy him a diamond ring in equal value to the engagement ring he had given her. Since Lucille was pregnant at the time of the marriage, this was probably a poorly hidden demand for reimbursement for the expense of the reluctantly purchased ring. Lucille refused. Sloan then demanded Lucille return the engagement ring he had given her. She did, then abandoned him for good. In a one-day divorce trial in Los Angeles Superior Court in September, Mulholland took the stand and testified that he told his daughter to return the ring, "as we did not care for any remembrance of him." Mulholland's imposing, patrician figure dominated the proceedings and Lucille was awarded her divorce, as well as full custody of baby Lillian; the details were read avidly by the citizens of Los Angeles in the morning papers.

The handsome Sloan proved worthy of the epithet "cad." His own mother, Clara S. Sloan of Pasadena, expressed her opinion of him in her will when she left him only ten dollars, stating that he had caused her much "heartache, disgrace, and humiliation." She instead left her $29,000 estate, including four land parcels in Fresno, Kern, and King's County, to her granddaughter, Lillian, who inherited it after Mrs. Sloan's death in August 1950.

Following the trial and, for once, keeping to the convention of the times, Lucille remained at the Mulholland home and seldom ventured out in public. To be a recent divorcée in 1918 was nothing to be flaunted, even for the daring Lucille. After patiently serving her sentence in the name of decorum, Lucille moved out of her father's house—and out of his control—to live on her own, leaving Lillian in the care of her former mother-in-law. Lucille went to work for a Pasadena law firm as a secretary where she met Benjamin C. Strang, the area's presiding justice of the peace. One year after her divorce from Sloan, Lucille wed Judge Strang, age thirty-seven, in a secret wedding at the lavish Pasadena home of Mrs. Waldo Falton. Following the late-night ceremony, the couple moved into a beautifully landscaped bungalow alongside the banks of the Arroyo Seco in Pasadena.

Strang had cleverly succeeded in having news of their marriage license suppressed, and they were married more than a week before it was made public. The first Mulholland learned of it was when news of the marriage appeared on the front page of the *Pasadena Star News* on October 30, 1919. In a diplomatic fashion more worthy of a presidential candidate, Mulholland calmly stated to journalists that he knew nothing of the matter at all, yet he wished his daughter and her husband every success and would immediately "send his paternal blessing." Picking up the event, Los Angeles newspapers, in deference to Mulholland's position as the leading city father, described Lucille's personality as "vivacious" and "possessing the independent nature that has made her father one of the great men of the Southwest."

While Mulholland may have told reporters he was delighted over his daughter's marriage, everyone close to him knew that he was less than pleased. The day after the announcement of Lucille's marriage, the heated Mulholland went to court escorted by his attorney, Lewis E. Whitehead, and filed a petition that he be named legal guardian of his two-year-old granddaughter, Lillian.

Certainly, the irresponsible Sloan rightfully deserved to be excluded from the Mulholland family as a ne'er-do-well. Lucille's selection of such a man could be excused as youthful impetuousness, and of course, there had been the pregnancy. But in Strang, Mulholland may have had more disturbing reasons to instigate his removal. Like Mulholland, Strang was a self-made man and successful in his own field of endeavor.

Benjamin C. Strang was born in Calhoun County, Iowa, where his Quaker father worked as a farmer. Strang was five years old when his family moved to Pasadena. As a child he worked as a newsboy, then during high school worked in the oil fields around Whittier, Los Angeles, and Kern County. Strang took up the study of law in 1912 and specialized in probate. In 1914, he was elected justice of the peace of Pasadena, and was reelected four years later. He was a staunch Republican and active in local politics.

The fifteen-year age difference that separated Strang and Lucille raised eyebrows. The very notion that an attractive, healthy young lady from a prominent family would involve herself intimately with an older man was distasteful, and unless something was direly lacking in her psychology, unexplainable. Such an alliance could only highlight Mulholland's failings as a father for his headstrong daughter. Conclusions could be drawn from the May-December romance that Lucille was seeking in Strang what she did not receive from her father—paternal love and understanding.

To the court, Mulholland swore that his daughter had no means with which to provide for Lillian's support and that the child was in the caring custody of its grandmother, Clara Sloan. Obviously, the responsibility of motherhood did not suit Lucille well during this young and confused period of her life, and Mulholland, unable to tend to the physical and emotional needs of the baby, proposed that the child remain with Mrs. Sloan; he would provide financially for her. On November 5, 1919, the court appointed him the child's official guardian. Sadly, Mulholland's shaky signature on court records attested both to his distress at the disheartening actions of his daughter and to his advancing affliction with Parkinson's disease.

On February 12, 1920, Lucille suddenly appeared at the home of

Clara Sloan and through "trick and device," as she was later charged, seized her baby. Clara Sloan immediately called the Pasadena police and charged Lucille and her new husband with kidnapping. With court papers drafted by her husband's law firm, Lucille avoided arrest and petitioned the court for revocation of Mulholland's guardianship, claiming that she had never given her consent. A court date was set for March 25.

News of the impending trial made the front page; predictably, Mulholland was livid. Clara Sloan filed a writ of habeas corpus against Lucille and Strang in an effort to have the child returned. Newspapers reported that the fight between father and daughter had grown ugly and spirited. At noon, March 26, deputy sheriffs armed with a writ of habeas corpus went to the Strang home to retrieve the child. Lucille barricaded herself inside and refused to turn Lillian over to the authorities. Newspapers reported events day by day until March 28, when Mulholland, embarrassed at the notoriety, wearily held a press conference and stated that he did not care to continue to air the family troubles in a public court proceeding, suggesting that he would not continue a strenuous fight against his daughter. On April 1, Judge Strang announced the case would be settled by all parties amicably out of court.

Mulholland's public pronouncements of reconciliation were deceptive. Father and daughter continued their heated arguments, until Mulholland managed to coerce Lucille into agreeing that Mulholland could keep Lillian at his St. Andrew's Place home, and both Lucille and Mrs. Sloan would be permitted rights of visitation. In exchange, Mrs. Sloan would withdraw the pending criminal charges. Four days later, Lucille had Strang deliver Lillian to her grandfather.

Ten months later, Lucille returned to her father's home, officially separated from her second husband, and by February 26, brought suit to have her marriage annulled. This time, a reconciled father and daughter, with Mulholland's attorneys, persuaded the court to exclude the public from the courtroom, and a subdued Lucille testified that her married life with Strang had been unhappy. After others testified on her behalf, Lucille meekly stated that her marriage to

Strang "had never been a reality." Again aided by the dramatic presence of her celebrated father in court, Lucille was granted an annulment on March 27, 1921.

Little is known about the reasons behind Lucille's breakup with Strang. One explanation is that the stress of the custody battle and the notoriety it generated destroyed much of the romance between the mismatched couple. Strang's sudden and unwanted fame in the papers may have caused him to consider terminating the marriage, and the constant pressure on Lucille from her father did little to reinforce any bond the lovers had between them.

Amazingly, less than three months later on Thursday, June 21, 1921, inside Mulholland's St. Andrew's Street home, Lucille was given away by her father—this time, with his blessing—to yet another husband, Ronald Robert Mack, a San Francisco stockbroker. Rose served as the maid of honor, and the groom's friend, Edgar Stone, served as best man. Only a few intimate friends attended the ceremony. The new couple then moved to Atherton, California, to start a home and to raise four-year-old Lillian.

Lucille's notorious affairs proved embarrassing, and her series of short-lived marriages had occurred during Mulholland's considerably rapid rise to national fame. The image-conscious Mulholland had worked vigorously to keep his personal life completely out of public scrutiny, yet Lucille's antics forced him to confront her through legal channels. For a man who loathed lawyers and bureaucracy, challenging his own daughter in a Los Angeles courtroom was especially repugnant.

Many of Lucille's friends claimed that her escapades were direct retaliations against the stern rule of her famous father; they were attempts to hit him where it would hurt the most—his pride. Another, perhaps truer, point of view was that the men Lucille married reflected key aspects of her father's complicated personality. In Edmund Sloan, Lucille found the rough-hewn field laborer, stubborn and unsympathetic. In the articulate Benjamin Strang, she found an older man with social prominence and prestige. In her third and final marriage, Lucille found a man in the middle of the spectrum, a respectful, caring, and successful businessman close to her own age and temperament, and acceptable to her father.

Sometimes in families, the most heated relationships are between members most alike. Like her father, Lucille was intelligent, stubborn, and fiercely determined. Like other children of American celebrities, she was unshielded from the crush of fame's weight. Had Mulholland been a more nurturing parent, she may have been inspired to dream great dreams and accomplish great deeds like her father. Mulholland was immensely successful but the price of his fame was borne by his children.

THE STORMY CONFLICTS between Mulholland and Lucille were not evident with his other children, yet his relationships with them were deficient, too. His third child, Thomas, never shared his father's commitment to hard work and was described as a self-centered loner, relishing obscurity like his mother. Never developing ambitions of his own, Thomas, like Rose, remained single. He never pursued a specific career and worked occasionally as his father's driver. Both he and Rose lived with Mulholland throughout their adult lives.

Studious and responsible Perry, the oldest son, married and raised a family. He worked his father's ranch for forty years, and grew to regret devoting his life to his father's ambitions instead of his own. Mulholland viewed Perry as the link to future Mulholland generations. While most of the sons of famous and wealthy Los Angeles men were sent away to Stanford or Harvard, Mulholland had other plans for Perry. In 1914, at age twenty-one, Perry was sent to farm the 640-acre Mulholland Ranch in the northwest corner of the San Fernando Valley, located between the tiny townships of Chatsworth and Zelzah (now Northridge). Mulholland began to purchase valley land one year before the finish of the aqueduct. For him, the land was an immigrant's dream of a comfortable future. But to Perry, it was a millstone.

Perry had hoped to pursue an education. He recalled his days in the White Mountains as a young man with the U.S. Geological Survey as the high point of his life. He had aspirations of a career in engineering or geology, but instead, spent the next forty years in subjugation to his father's dreams.

Perry's daughter Catherine wrote in her book, *Owensmouth Baby*, that during this period, a number of Los Angeles business and professional men bought ranching land in the valley and sent their reluctant sons to farm it. Mulholland did not purchase valley land for speculation, she wrote, but "with a landless Irish immigrant's dream of permanency," hoping that each of his five children would establish homes where he, the patriarch, "like Job would end his days, blest amidst his groves and heirs."

Perry went at his father's insistence to "join the ranks of young ranchers whose fathers were city men" to begin "a lonely bachelor's life in a prefabricated bungalow" in the isolated San Fernando Valley. In 1914, the valley was a sloping oval plain covering more than 100,000 fertile acres encircled by hills and mountain ranges. Because of the lack of water, the bulk of the valley was uninhabitable, and dry-farming crops like grain, hay, barley, and beans were all that were economically feasible before the aqueduct water arrived. Catherine Mulholland described her father's remote life in the valley during this period:

> To buy supplies, to eat an occasional restaurant meal and enjoy a little sociability, he had a Hobson's choice of the dinky settlements of Chatsworth and Zelzah or the newer little town of Owensmouth; the latter offered more amenities, however spare, and he went there with some regularity. He bought stock in the Bank of Owensmouth, obtained his groceries and meat from Mr. Sexsmith, one of the town's first grocers, sent his bean crops during the first World War to Vanomar Producers, the warehouse founded by Whitley and associates in Owensmouth, and ate an occasional chicken dinner at the Owensmouth Restaurant, run by Anna Gallow. He sometimes attended the high-stakes poker and gambling parties at the Alex Jeffrey and Brant ranches, and he went deer hunting with his good friend in Simi, David Strathearn.

In 1914, when aqueduct water was diverted for use by San Fernando Valley farmers, the tedious dry farming gave way to bountiful crops of oranges, lemons, apricots, peaches, plums, pears, nectarines, and winter tomatoes. Acres of fruit and nut groves multiplied across the valley, as did the human population.

Through a neighbor's matchmaking, Perry met Addie Haas, a third-generation San Fernando Valley daughter; they were married in October 1921 at the Owensmouth Haas ranch. Catherine Mulholland described the charming wedding of her parents:

> The groom's father, William Mulholland, enjoyed himself immensely, notwithstanding the fact that he'd absentmindedly put on a shirt with frayed cuffs which his dismayed daughters tried with manicure scissors to trim up. The unperturbed Chief waved them aside and became the life of the party. At one point, in high fettle, he complimented the bride's mother on the pink and white wedding cake by declaring that its smell reminded him of a girl he used to spark in Detroit.
>
> The groom's sister, Rose, was to have sung at the wedding. But at the last minute she'd dyed her hair to cover some premature grey and turned it such a jet black that she refused to stand before the assemblage and perform, so that there was no vocalist at the wedding, an omission which Addie never forgot—nor quite forgave.
>
> The bridegroom was late, having lost his cufflinks; he and his ranching neighbor and friend, Anstr Davidson, had had to make a bone-crunching run over dirt roads to Van Nuys to find another pair, and on the way had hit a chuckhole that almost dislocated both their backs. But the guests declared it a grand occasion, and after the festivities and a honeymoon in the Hawaiian Islands, the newlyweds returned to the Mulholland ranch where they would live for the next forty years.

In many respects, Perry Mulholland fulfilled the dreams of lineage of his immigrant father. He married a prominent Owensmouth lady, raised walnuts and oranges on what was now deemed ancestral Mulholland land, nurtured with water from Mulholland's own aqueduct system. He sired three grandchildren for his sainted father, two girls and a boy, and later cared for Mulholland the aging patriarch in his final years.

Denied the academic and professional life that he preferred, he served forth all of these things yet was never included in the personal and history-making affairs of his father. Mulholland instead preferred to keep daily company with Harvey Van Norman, and it was Van Norman, not the dutiful son, who shared the biggest part of Mulholland's life. No doubt Perry would have enjoyed the momentous events that Van Norman shared with his father, such as the many train trips to Washington and the glamour of state dinners and social events with nationally prominent politicians, world-famous authors, and businessmen. Perry must have felt envious of the young Van Norman, who basked in Mulholland's limelight, while he spent most of his adulthood as an isolated rancher toiling in his father's long shadow. No doubt by the time of his death in 1962, Perry Mulholland would wonder if it had all been worth it.

When Perry Mulholland first set foot on the ranch in 1914, his daughter Catherine wrote, he could see the dust from a car or wagon twelve miles away. By the time of his death, there were so many roads paved and so much traffic that he could scarcely maneuver a tractor from one grove to another. Perry's children would see homes that had held four generations bulldozed—the Mulholland ranch house itself was later replaced by a shopping mall and Kmart—and Perry's precious orchards paved over in asphalt, all the culmination of unprecedented, gluttonous growth after the aqueduct water arrived. But during this period from 1914 to 1924, the San Fernando Valley prospered as a booming pastoral community with unblurred visions of a certain, bounteous future.

9

EDEN

Thou shalt be like a watered garden,
and like a spring of water,
whose waters fail not.
ISA. 54:11

MULHOLLAND AND CITY LEADERS were certain that the aqueduct would bring into Los Angeles eight times as much water as it needed and four times as much as Los Angeles could ever use, and any valley community might take all the water it needed—provided that it became part of the city. Soon the greater portion of the San Fernando Valley, almost 200,000 acres (275 square miles) was annexed to the city of Los Angeles. Irrigated acreage in the San Fernando Valley rose from 1,000 acres in 1913 to 75,000 acres in 1918, as the same river that kept the Owens Valley green now turned the San Fernando Valley into a garden.

Lured by the promise of cheap land, good wages, and plenty of water, the San Fernando Valley population expanded rapidly. Enticed by Otis and Harriman's advertising in national magazines and newspapers, thousands of people flocked to the valley. They arrived on E. H. Harriman's Union Pacific Railroad, commuted on Sherman and Huntington's trolley cars, and read Otis's newspapers. Settling in the San Fernando Valley, a Harriman-Sherman-Huntington-Otis development, they drank Mulholland's Owens River water.

Otis, Chandler, Whitley, Sherman, Brant and the rest of the Board of Control presided over much of the valley's metamorphosis. In one of the syndicate's first meetings, Harrison Gray Otis announced the group's intention. "We do not want to sell a town site. We want to build a town." Between 1911 and 1915, the Board of Control created multiple new town sites from their vast holdings; true to their self-serving ways, the board named these new towns after their own. Marion, known today as Reseda, was named in honor of the daughter of Harrison Gray Otis and wife of Harry Chandler. The town of Van Nuys was named in honor of Harry Chandler's old friend Isaac Van Nuys.

The syndicate agreed, however, not to name the new towns after themselves, following the public clamor concerning their alleged aqueduct conspiracy. But eventually the town of Sherman Oaks materialized, named after the balding ex-school administrator who had become a trolley magnate. The first highway to intersect the valley was also dubbed Sherman Way; and a prominent street in Van Nuys was named Hazeltine, after one of Sherman's daughters. However, Owensmouth, later renamed Canoga Park, was named for its proximity to the terminus of Mulholland's aqueduct. The men in the syndicate were happily engaged in the "drama of city-building."

Harrison Gray Otis, with his huge belly, walrus-like mustache, and goatee, provided much of the booster rhetoric concerning the syndicate's elaborate plans and signed most of the checks. After incorporating as the Los Angeles Suburban Homes Company in 1909, the five energetic capitalists intended to profit handsomely from their master-work of three new towns complete with roads, schools, utilities and water services, and modern transportation. When the Board of Control had finished naming boulevards and town sites, the members began to carve out choice tracts for themselves, like feudal princes dividing a conquered territory. General Sherman allocated 1,000 acres at the site now known as Sherman Oaks; Otis took 550 acres that he later sold to author Edgar Rice Burroughs, who renamed it Tarzana; Brant appropriated 850 acres to build his stunning Brant Rancho; and Whitley and Chandler selected handsome tracts for themselves near Sherman Way and Van Nuys Boulevard.

Catherine Mulholland described the land syndicate:

> They shared the belief derived from Calvinism that material success is outward evidence of divine election, so wealth was indeed a blessing. . . . They spoke of vision, progress, and of upbuilding Southern California. They adored the game of Profit and Loss, none more so than the man who was to be their field marshal in the new Valley campaign—Hobart Johnstone Whitley, the Great Developer.

H. J. Whitley was the primary creator of the new San Fernando Valley. A six-foot-tall, three-hundred-pound, square-jawed, grizzly-looking man, Whitley was described as a "six cylinder working man of electric optimism." Credited with the creation of Hollywood, then one of the prettiest suburbs in Los Angeles, Whitley was known for his extraordinary town-building in Oklahoma and other states in the Midwest. Once the Pacific Electric trolley car system was extended into the valley, Whitley planned subdivisions. With the arrival of the Red Cars, the valley was now within an hour's ride on the trolley from downtown Los Angeles, and no longer an isolated wasteland.

Whitley's bold plans for the valley boulevard known today as Sherman Way, was vividly described in Los Angeles's *Sunset* magazine in 1914:

> They gave him carte blanche. He built the boulevard. He made it two hundred feet wide, with the trolley in the center and a driveway on either side. He laid an oil-macadam pavement for general

traffic and an asphalt concrete pavement for exclusive automobile traffic, built concrete curbs for sixteen miles of the twenty-two miles now built. He laid out a strip of parking along the trolley tracks, other fifty foot strips of parking on either side. Then he proceeded to embellish the highway.

First he planted rosebushes five feet apart, four rows of them on the main boulevard, two rows on the connecting links, fifty miles of roses in all. Not the ordinary hedge variety but blooded stock, American beauties, La France, Cécile Brünner, tea roses, roses of all colors and hues, roses the size of soup plates and roses just right for a boutonniere. Behind the roses he planted rows of exotic flowering shrubs, oleanders red and white, ornamental shrubs that turn themselves into masses of flaming yellow, of royal purple and vivid crimson when the spirit and the season prompt them.

Behind the bushes he set out a row of magnolias from the South, flowering acacias from Australia, alternating with stately fan palms from the Canaries. Next he jumped to India, to the shoulders of the Himalayas, planted a double row of the deodars made famous by Rudyard Kipling. Behind the silvery gray foliage of the deodars he found room for more ornamental shrubs, and at the outer edges of the broad parkings he supplied somber Monterey Pines as a fitting background. Then he rested.

If there is anywhere a highway that exceeds Whitley's twenty-two-mile boulevard in width, length and in the variety and character of its arboreal ornamentation it has succeeded effectively in hiding its light under a bushel basket.

It was Whitley who faced the daily logistical problems of road building and utilities and staging the elaborate sales campaigns that were used to lure prospective home buyers to the desolate, arid valley. He hosted "two years of barbecues, Rose festivals, Poppy Days, car races, and ballyhoo" to sell the syndicate's Owensmouth subdivision. The new San Fernando towns were ingeniously promoted and sold to Americans hungry for a better life in the promised land of southern California where all things were possible. The creative sales pitch used by Whitley and others were described by Kevin Starr in *Material Dreams:*

> The selling of these homes, all 300,000 plus of them, involved flamboyance, gross exaggeration occasionally and sometimes deliberate deception. . . . Perhaps the most fantastic of all subdivision concepts and sales pitches involved Girard in the western San Fernando Valley, a Potemkin village of false fronts held up by the rear braces so as to suggest the city would soon rise there.
>
> A platoon of salesmen, warmed earlier to the task by group calisthenics, met prospects arriving by bus. A numbered name tag affixed to his or her lapel, each prospective buyer was then led by a single salesman, drawn by lottery, through a rehearsed sequence of lunch followed by a walking tour of the subdivision. At the right moment the salesman brought the client to a specially placed closing booth, where a senior sales specialist cinched the sale.

Through Whitley's superb salesmanship, the Board of Control's 1911 purchase of the 47,500-acre Porter ranch grew into one of the biggest subdivisions in the world. No one knows exactly how much profit the San Fernando land syndicate realized from their initial investment, but writer William Kahrl estimated that Harry Chandler was worth as much as $500 million when he died in 1944. H. J. Whitley, however, was not so lucky. After a series of disastrous investments he lost nearly everything.

The land boom in the San Fernando Valley generated by the Board of Control following the delivery of aqueduct water has been compared to a Louisiana Purchase for the city of Los Angeles. "Because the oligarchy controlled or decisively influenced the governmental bodies which controlled the water, it grew rich and powerful beyond measure," noted Kevin Starr. "It also helped that the most active land speculator owned the dominant newspaper. Thus one might make reference, almost, to a Southern California Raj—an orchestration of business, financial, political and government power, all of it controlled by one oligarchy. . . ."

Mulholland always insisted that he was never involved in land speculation for profit, and had no ulterior motive in the aqueduct's construction. When one reporter charged him with collusion in the land syndicate, Mulholland erupted in anger.

"Arable lands which should be selling at about $100 an acre have been seized by a few capitalists who have forced prices to $1,000 an acre," Mulholland countered, segregating his "duty as an engineer" from the scheme of land speculators. "Instead of being developed as agricultural lands, the property has been subdivided into town lots and small 'rich men's country estates' at prohibitive values. The men who bought up this property have looked forward to the time when the aqueduct would be completed and the plans for distribution of the water through this territory would enhance land values." Mulholland's verbal attacks were directed at the Board of Control, and were part of his continuing efforts to separate himself from them at all cost. Indeed, he felt it was critical to distance himself from the land syndicate and he did so by refusing to purchase tracts from or near the men under attack.

Others in Mulholland's circle bought land in the valley as well. Harvey Van Norman bought hundreds of acres near the Mulholland parcel, and kept it a secret. Other men from the Department of Water and Power bought adjoining land in Chatsworth, including J. B. Lippincott, Ezra Scattergood (chief engineer of electrical division), Phil Wintz (Haiwee Dam engineer), Roderick McKay (an aqueduct engineer whose daughter became the only woman rancher in the area), and Eva H. Shoemaker, Mulholland's only female secretary. Despite his financial interest in the San Fernando Valley land he owned, the facts confirm that Mulholland was sympathetic to the agricultural interests in the valley as opposed to the interests of the land developers.

DESPITE HIS DENIALS of profiteering and his pontifications on the qualities of the working man, Mulholland was now a wealthy man himself, and by the time of his death had accumulated an estate valued at over $700,000, enormous by 1930s standards. The thrifty lifestyle that he and his family maintained, combined with shrewd real-estate investments, contributed to the large fortune. Aside from employing a full-time car and driver, Mulholland lived a middle-class life in a modest four-bedroom house; he wore store-bought clothes and ate average meals. His public-benefactor image was further enhanced by his sedate lifestyle, and good, quiet, Christian living.

The bulk of Mulholland's estate was derived from the San Fernando Valley lands he had acquired from 1912 to 1919 at prices between $50 and $150 an acre. The 640-acre ranch that Perry Mulholland cultivated for his father was later valued at $511,315, a huge return on his investment. Mulholland never earned the tens of millions of dollars realized by some members of the Board of Control, and Mulholland was not guilty of profiteering on that scale. He did, however, own land in the San Fernando Valley that he bought for bargain prices based on its limited dry farming yield before the aqueduct water arrived.

"He is unquestionably honest," said the comptroller of the Department of Water and Power, in defense of Mulholland's character. "Although he is a wealthy man, he has never been avaricious, but rather has made his money by investing wisely in well-selected real estate. He is a careful liver, has no extravagant habits, and has never made excessive charges for his services."

Mulholland could also boast a remarkable salary. His $10,000 salary as chief engineer for the Department of Water and Power was the highest paid to any civil servant in Los Angeles, and since 1886, he had drawn a higher salary than the mayor, district attorney, or chief of police. Later in his career he earned as much as $15,000 a year as chief of the water system.

Mulholland's financial management showed common sense and frugality, but he did have one extravagant quirk; he was known for giving expensive, unexpected gifts to friends and associates, generally bought impulsively, paid for in cash, and delivered in surprise by a third party. He also lavished expensive gifts on his family. Friends and associates said that in his personal life, Mulholland seemed rather careless with money, failing to deposit his paychecks or leave money for the family to purchase groceries.

In addition to his salary, Mulholland engaged in lucrative consulting contracts with other cities, serving as an expert in the development of public water projects for San Francisco, Oakland, and Seattle. He also consulted for corporations such as the Western Union Oil Company, where his reports about oil formations resulted in the discovery of productive oil fields that created fortunes in San Luis Obispo County. He did this consulting work throughout his career, as early as 1899, and collected ever-larger fees. During the aqueduct's construction, however, he turned down the work because of the demands on his time.

Despite his substantial portfolio of real estate, stocks, bonds, and leaseholds, Mulholland never exhibited outward signs of his personal wealth, always remaining loyal to his modest immigrant roots.

IN MARCH 1920, at age forty-five, Joe Desmond, the food contractor on the aqueduct, died, and news of his death brought Mulholland together with Raymond Taylor. The death came as a shock to all who knew him, and Desmond's widow, Alice, held funeral services at the large Desmond home and later at the Cathedral Chapel. Mulholland and Taylor served as pallbearers.

Desmond had taken much of the brunt of the fraud charges during the Aqueduct Investigation Board's inquiry in 1912. Taylor had grown fond of the ambitious, flamboyant Desmond during their years together on the aqueduct line despite their differences. After the aqueduct was finished, Desmond had garnered a lucrative twenty-year contract as the hotel and camp concessionaire in Yosemite National Park, where he had formed the Desmond Park Service Company and amassed further wealth.

Daniel "Joe" Desmond, a member of one of the oldest families of the city, was born in 1875 at the old family residence on the site now occupied by the Chamber of Commerce building. He attended public schools and then graduated from St. Vincent's College. As a young man, Desmond had worked as the head of the relief committee sent to San Francisco to feed the homeless following the earthquake and fire in 1906. It was his admirable work there that brought him to the attention of the Bureau of Public Works and General Chaffee, who retained his services as commissary chief for Mulholland's aqueduct.

Standing at the grave of his former friend and watching his elaborate flower-draped coffin being lowered into the ground, Dr. Taylor recalled a conversation he had shared with General Adna R. Chaffee and Joe Desmond several years earlier. By then, both of the men's connection to the aqueduct had long since been concluded.

Chaffee had turned to Taylor and asked, "Doctor, how much money did you make off this job?"

"General, I can't tell you exactly, but I think I made about $6,000 a year for five years; in other words, about $30,000 in the clear," Taylor replied.

Chaffee then turned to Desmond and asked, "How about you, Joe, how much did you make?"

"About a quarter of a million," said Desmond matter-of-factly.

The general slammed his fist on the table and said, "Damn it Joe, you made too much and the doctor didn't make enough!"

When Taylor later recounted the story to Mulholland, Mulholland was shocked to learn that Joe Desmond had made such a fortune. Later, Joe Desmond Jr., Desmond's only child, inherited a substantial estate, a large part of it proceeds from the Los Angeles aqueduct contract.

Others besides Desmond profited handsomely from the aqueduct, but Mulholland had devoted much effort to ensuring that the enterprise was built solely for the enrichment of the public. The notion that a man should be compensated fairly for his efforts on behalf of the city, but not made rich, was most apparent in Mulholland's dealings with Fred Eaton and the endless haggling over the price for Long Valley. Even when it made the most sense to buy the ten thousand acres, Mulholland refused to do it—never giving in to what he perceived as blackmail against the people. Learning about the fortune Desmond had garnered during his years as concessionaire on the line troubled Mulholland.

During the services, Dr. Taylor thought back to the days when he and Desmond engaged in heated arguments over the food served to the men on the aqueduct, yet he remembered one special dinner Desmond's cook had served him, complete with a bottle of expensive wine. Taylor and Desmond were professional colleagues, and had shared hardships, but they were never close friends. Like many others, what they had in common was their appreciation for William Mulholland and their big effort in the desert.

10

WARS IN HEAVEN

Whence come wars
and fighting among you?
JAMES 4:1

JOE DESMOND'S FUNERAL had reinforced Mulholland's own sense of mortality. Mulholland was now sixty-five years old, and the city of Los Angeles had changed drastically since his rise to superintendent of the Los Angeles City Water Company in 1886. The population of Los Angeles was now diverse in its ethnicity; its politics had become complex and the direction of its growth unclear.

By 1920, the city's population exceeded half a million. The movie industry was thriving, oil production was booming, and Los Angeles's harbor imports surpassed those of San Francisco. Los Angeles real estate became a principal commodity, and in the next eight years, over 3,000 subdivisions encompassing 50,000 acres were formed and 250,000 building lots created.

Land developers and publicists went to work informing the world that "God made southern California—and made it on purpose." Harry Chandler's All Year Club lured millions from the Midwest, where thousands of real-estate agents sold them land. By 1920, Santa Monica Boulevard, once a dry and dusty strip, was an elegant, palm-lined corridor. Small towns, orange groves, shops, and industries sprang up where before there had been empty expanses of dusty savannah.

The movie industry moved West and Hollywood's outdoor movie sets resembled African jungles while smart, city bungalows featured thick, green lawns. Nighttime Los Angeles was now a "wonderland of light" and "in sheer extent it was a horizontal equivalent of vertical New York"; no other spectacle like it existed anywhere in the world.

Tourists could ride a mountain trolley sixty-one hundred feet into the air to the top of Mt. Lowe and behold the staggering Los Angeles basin, Pasadena, and fifty-six contiguous cities and suburbs "spread out over a vast sea of illumination." To anyone who saw it, the vista was "vaguely fascinating, beautiful almost, in that confusion of lights, in those flashing signs and advertisements, in those streams and rivulets of motor headlights on the boulevards, a rhythm in the distant roar of the city . . . fantastic long slender shafts of light. . . ." However, the unparalleled growth was not without its detractors. "The community is thus a parasite upon the great industrial centers of other parts of America," wrote Upton Sinclair. "It is smug and self-satisfied making the sacredness of property the first and last article of its creed. . . . Its social life is display, its intellectual life is 'boosting' and its politics are run by Chambers of Commerce and Real Estate Exchanges."

In the six years since Mulholland had finished the aqueduct, Los Angeles had undergone a staggering transformation; its population was soon to double the pre-aqueduct figures. The furious growth was severely straining the new water system's capacity. Though brilliantly designed, it had one great defect—a lack of storage reservoirs to maintain a predictable flow in drought years. By his own admission, Mulholland had failed to predict the unparalleled growth of the city, and the growth upset the calculations on which he had based his plans for the Los Angeles Aqueduct. "If I had [had] more faith in the growth of the city," Mulholland said, "I would be better off today." Relying solely on the Sierra Nevada snowpack for a constant stream of water, the aqueduct was useless in times of severe drought and by the summer of 1923, after a long dry spell, the voracious needs of San Fernando Valley farmers had literally exhausted the aqueduct's resources. A mere ten years since the big pipe's construction, the aqueduct could no longer serve both urban Los Angeles and the agricultural empire now annexed in the San Fernando Valley.

THE WORSENING SCENARIO confronting the city's water supply did not go unnoticed by Fred Eaton, who knew all along this day would come. Eaton had no doubt that in time, he would receive his full price for Long Valley, and to ensure that end, endeavored to protect himself legally. In 1923 he sought an injunction to prevent the construction by the city of Los Angeles of a proposed 150-foot dam at Long Valley. Eaton's shrewd easement to the city, finalized in 1905, had permitted a dam only 100 feet high, severely limiting its usefulness as a northern storage facility for the Los Angeles Department of Water and Power. Fearing the city would exceed its legal rights and build the 150-foot dam anyway, Eaton filed a complaint on July 30 in Mono County against the City of Los Angeles, enjoining it from construction.

Without Eaton's Long Valley site to regulate the flow, Mulholland knew Los Angeles was doomed to recurring water shortages. It was also now apparent that as the city was forced to take more and more water, the Owens Valley would slowly die. Mulholland pleaded with Eaton to sell the parcel for less than a million dollars. Eaton refused. This time, Mulholland announced that he was severing all ties with his former friend. "I'll buy the Long Valley three years after Eaton is dead," a scowling Mulholland told Van Norman. It was a promise eerily close to prophecy.

AFTER THE DEFEAT of the "socialist-inspired" attacks and exoneration by the Aqueduct Investigation Board in 1912, the war waged against Mulholland started again. This time, however, the battle was not a propaganda tool by "embittered socialists" to win an election, but was a twenty-year war over water, instigated by Owens Valley ranchers.

William Mulholland admitted openly that he "savored the combat" of building the great aqueduct. Verbal attacks against the aqueduct by the people of the Owens Valley had begun in 1904, when gossip spread throughout the valley about Mulholland's plans and citizens were quick to react to the threat it posed to their valley and the Federal Reclamation plan on which they had pinned their hopes. The 380 vociferous signatories who controlled 104,242 acres of the Owens Valley petitioned the Secretary of the Interior to urge that the federal irrigation project be continued.

Mary Austin, who later left the Owens Valley for a celebrated literary career, was convinced that the valley had died when it sold its first water rights to Los Angeles—and that the city would never stop until it devoured every drop of Owens River water. Many had feared Austin's analysis was right.

In 1905, a farm woman from the small town of Poleta, fifteen miles east of Bishop, wrote a vivid letter to President Theodore Roosevelt that expressed the heartbreak and trepidation of the Owens people:

Pres. Theodore Roosevelt
Washington, D.C.

Dear Friend:

Look on your map of California, along the eastern boundary south of Lake Tahoe and you will find a county named "Inyo." Running into this county from Nevada through a small corner of Mono County you will see the Carson and Colorado railroad which after it enters Inyo follows along the Owens River until they both come to Owens Lake, an alkaline body of water. It is about this river I write to you.

This river after it leaves the narrow mountain canyon, runs through a broad and fertile valley for 100 miles. The first 20 miles of which is all or nearly so, in cultivation, further south ranches become more scattern. It has four prosperous towns.

Indeed the people are very proud of their little valley and what their hard labor has made it. The towns are all kept up by the surrounding farms. Alfalfa is the principle crop. They put up to from two to four ton per acre and it cost from $1.25 to $1.75 to put it up and sell for $4 to $7 per ton, so you see the county is very prosperous. As there is about 200,000 ton raised in the valley if not more every year. Cattle raising is a great industry.

There has never been any capitalist or rich people come here until lately and all the farms of the Owens valley show the hard labor and toll of people who came here with out much more than their clothes. And many had few of them.

Now my real reason for wiring this is to tell you that some rich men got the government or "uncle Sam" to hire a man named J. B. Lippincott to represent to the people that he was going to put in a large dam in what is known as Long Valley. But—Lo! and Behold! Imagine the shock the people felt when they learned that Uncle Sam was paying Mr. Lippincott. He was a traitor to the people and was working for a millionaire company. The real reason for so much work was because a man named Eaton and a few more equally low, sneaking rich men wanted to get controlling interest of the water by buying out a few or all of those who owned much water and simply "freeze out" those who hadn't much and tell them to "Git."

Now as President of the U.S. do you think that is right? And is there no way by which our dear valley and our homes can be saved? Is there no way by which 800 or 900 homes can be saved? Is there no way to keep the capitalists from forcing the people to give up their water right and letting the now beautiful alfalfa fields dry up and return to a barren desert waist?

Is there no way to stop this thievery? As you have proven to be the president for the people and not the rich, I, an old resident who was raised here appeal to you for help and advice.

My husband and I within the last year have bought us a home and are paying for it in hard labor and economy. So I can tell you it will be hard to have those rich men say "stay there and starve" or "Go." Where if we keep the water in the valley it won't be only three years until the place will pay for itself.

So help the people of the Owens valley!

I appeal to you in the name of the Flag, The Glorious Stars and Stripes,

Yours Unto Eternity,
Lesta V. Parker

Lesta Parker's troubles may have seemed trivial against the weight of larger looming political conundrums, and President Roosevelt's own ambitions and political strategy as it concerned California dictated an administration policy of massive public works to benefit urbanization. "Booming Los Angeles with its municipal water department was a trophy that the hunter's eye did not miss," and the eventual demise of the Owens Valley, according to historian John Walton, fell neatly into a groove made for it by the transformation of the state in the Progressive Era.

Protests in the valley were more silent once Mulholland began construction of the aqueduct in 1908. Some aqueduct camps suffered occasional arsonist attacks, but Mulholland was never certain if they had originated from labor disputes or citizen sabotage. In general, up to 1920 Los Angeles took only excess surface water for the aqueduct, and agriculture in the Owens Valley thrived on booming prices during World War I. Negotiations for valley underground storage water continued from 1913 through 1927, but no agreement was ever finalized.

In 1919, a series of crippling droughts forced the city to tap Owens Valley ground water and pump it into the aqueduct, depleting the valley's water table, and the once-lush Owens farmland began to dry up. Forceful criticism against the city began to surface, and even Will Rogers informed the nation: "Ten years ago this was a wonderful valley with one-quarter of a million acres of fruit and alfalfa. But Los Angeles had to have more water for its Chamber of Commerce to drink more toasts to its growth, more water to dilute its orange juice and more water for its geraniums to delight the tourists, while the giant cottonwoods here died. So, now this is a valley of desolation." What was feared had finally occurred, and families were forced to shut down businesses and farms. The once sprawling Owens Valley was being rapidly transformed into a dust-laden wasteland.

In the face of new uncertainties, both sides adopted more aggressive strategies. Los Angeles began wholesale purchases of land and water rights, and city property acquisitions soared from six per year to over one hundred by 1923. The city used a checkerboard pattern in land purchases, focusing on strategic parcels, leaving others cut off from water sources. Owens Valley property owners and canal companies united in a single association called the Owens Valley Irrigation District to stop Los Angeles city crews from diverting water away from their property. In August 1923, city crews continued reclaiming water for the aqueduct by breaking locally owned dams and canal heads, and early in 1924, Los Angeles sought an injunction in the Inyo Superior Court to prevent the sale of irrigation bonds to finance the Owens Valley Irrigation District. Anger over the city's efforts to undermine the irrigation district exploded in violence, and what was to be called the "California Water Wars" began.

IN THE EARLY HOURS of Wednesday, May 21, 1924, the Los Angeles Aqueduct was dynamited just below the Alabama Gate spillway. The force of two hundred pounds of dynamite lifted a huge boulder alongside the channel, but most of the debris fell back in the hole. The blast was less effective than planned, causing twenty-five thousand dollars' worth of damage. Mulholland dispatched maintenance crews that repaired the hole in two days.

The desert sand surrounding the scene of the crime was covered in tire tracks and footprints indicating that as many as fifty men had participated in the early morning raid. The dynamite was believed to have come from a warehouse owned by two brothers involved in Inyo banking, Wilfred and Mark Watterson. Witnesses reported seeing a caravan of eleven automobiles traveling from Bishop, later passing through Big Pine, gathering supporters en route.

Fearing local law enforcement to be sympathetic to the saboteurs, Mulholland dispatched private investigators to report on the incident. Private eye Jack Dymond, hired by the Los Angeles district attorney's office, said that local sentiment was obstructing his investigation, and predicted that the criminals would never be caught because no local grand jury would indict, and no court would convict them. The city offered a ten-thousand-dollar reward for any information leading to conviction. The insurrection had interrupted interstate telephone communications, a federal violation, and city officials hoped the Federal Bureau of Investigation would step in.

Various theories about the responsible parties covered the pages of Los Angeles newspapers. The *Times* endorsed a theory that a "known red leader" fired from the aqueduct was behind the takeover, and privately, water officials suspected "spite worker" involvement based on the dynamiter's intimate knowledge of the aqueduct. The *Los Angeles Examiner* described the culprits as "maniacs and anarchists," urging the District Attorney's office to dispatch an anti-Wobbly squad to bring the mob to justice.

ASA KEYES (rhymes with "tries"), the city of Los Angeles's young, ambitious, and colorful district attorney, quickly entered the case. As a politician of shrewdness and skill, the outspoken Keyes recognized immediately that public outrage over the bombings could build his own career, and he pounced on the occasion, calling for the immediate arrest and indictment of the dynamiters.

But Asa Keyes's hot pursuit of the "anarchists" faded as his investigation ended in no arrests or indictments. In one heated press conference, Keyes fumed that every resident in the valley "knew damn well" who did the dynamiting, but no one would say.

There was more violence. The first bombings in May 1924, were random, disorganized affairs that did little damage, yet caught the eye of the national press and infuriated Mulholland. The Lone Pine Canal was bombed a few weeks later when twenty carloads of ranchers detonated three boxes of powder, breaching the canal wall. The warning was clear.

Over the next six months, William Mulholland received hundreds of death threats, delivered by telephone and in the mail, both at home and at the office, intimating that he would be killed if he visited the Owens district. Mulholland never appeared concerned, announcing defiantly that despite the threats against his life, he would accompany the president of the Los Angeles Chamber of Commerce during his visit to the valley the following week.

In fact, Bill Mulholland probably did not take the threats against his life seriously or fear for his safety. But it is clear that he greatly underestimated the antagonism in the valley and the violent lengths to which the men would go to further their cause.

Led by brothers Wilfred and Mark Watterson, an organized Owens Valley citizenry posed a formidable threat to the new water system. The Wattersons' Inyo County Bank was the region's primary source of economic life. During the post-war recession of the early 1920s the brothers had gained control of other banks in the region and refinanced most of the valley's farms and ranches. Although they were trusted by the Owens Valley people, who often boasted they never foreclosed on a valley farm or sued a valley debtor, many ranchers and farmers were beholden to the brothers because of the mortgages they held.

Meanwhile, city agents continued to checkerboard the valley by purchasing nonadjacent plots of land and urging holdouts to sell. Their patience and nerves worn out, and their very existence now in imminent danger of extinction, the ranchers reacted violently. Lynch mobs grew in size, and roamed the territory at night brandishing nooses before terrorized valley residents who dared to show signs of selling out. In Bishop, farmers forcibly withheld water from the streams that fed the aqueduct. These acts of terrorism at first were ignored by Los Angeles city officials, unaware that they were indicative of a collective valley sentiment of rage.

Frustrated, Mulholland urged Harvey Van Norman and W. B. Mathews to go to the Owens Valley and do what they could to strike a deal. Led by the Watterson brothers, distrust was so widespread that every city offer was met with a warning that unless handsome reparations were guaranteed, the valley would not deal. It appeared the valley was not interested in a treaty, only in money, and hardened its position.

Upon his return, Van Norman reported that he found Mulholland unwilling to listen to the many details concerning the city's negotiations with the Owens Valley, and that Mulholland was distracted and aloof regarding the matters in Inyo County. His mind was now fixed on the problem of obtaining new water resources for the city and he left the entire Owens Valley matter in the hands of Van Norman and Mathews, who faced a seemingly endless series of clashes with valley ranchers. Every time it appeared a settlement might be reached it failed, and the water war assumed a pattern of alternating violence and negotiation.

Mulholland's pugnacious, difficult personality, the same no-nonsense stubbornness that created admiration among his subordinates, worked to his disadvantage when confronted with the seething sentiments against him from the Owens Valley. When a reporter from the *Times* asked him why there was so much dissatisfaction in the Owens Valley, Mulholland quickly retorted, "Dissatisfaction in the valley? . . . Dissatisfaction is a sort of condition that prevails there, like foot-and-mouth disease." It was the same unreasoning rage that made him say, according to author Marc Reisner, when Mulholland's war of attrition against the Owens Valley had finally caused events to take a drastic turn for the worse, that he "half-regretted the demise of so many of the valley's orchard trees, because now there were no longer enough live trees to hang all the troublemakers who lived there."

ALREADY PAST RETIREMENT AGE, bored with the never-ending and tedious Inyo County war, unnerved by his aqueduct's failure to keep pace with the city's growth, and perhaps fearing forced idleness, Mulholland was already dreaming about a last great achievement. He would guide, at least through its preliminary stages, a final great water project on behalf of his beloved city—the giant aqueduct from the Colorado River, and he used his considerable influence to further that cause.

"There are no more streams, small or large, within the boundaries of the state which may be developed by the great cities of southern California," Mulholland told reporters. "Fortunately, however, California is bordered on the east by one of the greatest rivers in America, the Colorado."

Mulholland's continual, obsessive quest for water at times manifested itself in outrageous notions—proposals which seemed almost out of step with sanity. Horace Albright, later director for the National Park Service under Herbert Hoover, recalled meeting Mulholland at a testimonial dinner in honor of Senator Frank Flint, and his surprise at Mulholland's bizarre solution for a big dam in the state of California.

At the time, Albright was a young park superintendent seated at Mulholland's table. Midway through the evening's program, Albright felt Mulholland tap him on the shoulder. Albright's recollections were made to Marc Reisner, author of *Cadillac Desert.*

"You're from the Park Service, aren't you?" Mulholland demanded, more than asked.

"Yes, I am," said Albright. "Why do you ask?"

"Why?" Mulholland said archly. "'Why?' I'll tell you why. You have a beautiful park up north. A majestic park. Yosemite Park, it's called. You've been there, have you?"

Albright said he had; he was the park's superintendent.

"Well, I'm going to tell you what I'd do with your park. Do you want to know what I would do?"

Albright said that he did.

"Well, I'll tell you. You know this new photographic process they've invented? It's called Pathe. It makes everything seem lifelike. The hues and coloration are magnificent. Well, then, what I would do, if I were custodian of your park, is I'd hire a dozen of the best photographers in the world. I'd build them cabins in Yosemite Valley and pay them something and give them all the fill they wanted. I'd say, 'This park is yours. It's yours for one year. I want you to take photographs in every season. I want you to capture all the colors, all the waterfalls, all the snow, and all the majesty. I especially want you to take photographs in every season. I especially want you to photograph the rivers. In the early summer, when the Merced River roars, I want to see that.' And then I'd leave them be, and in a year I'd come back, and take their film, and send it out and have it developed and treated by Pathe. And then I would print the pictures in thousands of books and send them to every library. I would urge every magazine in the country to print them and tell every American and museum to hang them. I would make certain that every American saw them. And then," Mulholland said slowly, with what Albright remembered as a vulpine grin, "and then do you know what I would do? I'd go in there and build a dam from one side of that valley to the other and *stop the goddamned waste!*"

"It was the tone of his voice that surprised me," Albright told Reisner. "The laughingly arrogant tone. I don't think he was joking, you see. He was absolutely convinced that building a dam in Yosemite Valley was the proper thing to do. We had few big dams in California then. There were hundreds of other sites, and there were bigger rivers than the Merced. But he seemed to want to shake things up, to outrage me. He almost *wanted* to destroy."

In 1923 the city suffered another period of drought and Mulholland publicly announced a more realistic proposal and the need for a "vastly greater water supply." In October, voters in Los Angeles approved bonds to give Mulholland funds to study the possibilities, and in December, he launched a survey to determine the most practical route for an aqueduct from the Colorado River to Los Angeles. Mulholland, Van Norman and W. B. Mathews, among others, ventured on a ten-day float in four rowboats through the majestic river where the party scrutinized 150 miles of river and terrain in a "brush-clearing expedition."

One famous press photo, depicting Mulholland and Van Norman aboard a Colorado River rowboat, published widely in California newspapers, continued to instill in the public mind the Mulholland mystique and helped persuade the city that water from the Colorado River was needed to support Los Angeles's projected population of 7.5 million people. In sharp contrast to the rough survey days in 1905, Mulholland and his engineers would sleep in hotels and eat in restaurants; yet Mulholland hoisted his sleeping bag on his back and posed for newspaper photographers, in a deliberate promotion of the "Mulholland mystique."

It was in many ways a recapitulation of Mulholland's buckboard-and-whiskey journey with Eaton in 1904, but this time the scale of what was envisioned increased fourfold. Flowing at 400 second-feet, the Owens River served only 2 million people, with 1500 second-feet capacity. The Colorado River could provide for 7.5 million—a huge billion-gallon-a-day water carrier. Like his earlier trip with Eaton, Mulholland and his colleagues also shared a deep sense of change and forced destiny as the men discussed the ingenious political and engineering maneuvers that were required to harness this new water source.

Over the next four years, Mulholland sent sixteen survey parties to Colorado to study 60,000 square miles and five possible routes for the Colorado Aqueduct. "The preliminary work will require a year and half to two years," Mulholland told reporters, but it would be an "incomparably simpler task" than planning the route of the Owens River aqueduct, which operated by gravity alone, and required an enormous number of cuts and tunnels. But "those were the days before people took to roosting on the hills like turkey cocks," Mulholland mused. Now the elevated Los Angeles suburbs would require the delivery of Colorado River water by pumping.

Mulholland's survey teams considered dozens of possible routes for the Colorado River aqueduct, and submitted detailed maps of the five most suitable proposals. The most favored route with only minor changes was exactly the route Mulholland had sketched, completely without the aid of instruments on the "brush-clearing expedition" with Van Norman and Mathews in 1923, an incontestable example of Mulholland's extraordinary ability.

Mulholland's perpetual search for water made him a pawn to those groups he detested most, the money makers and the politicians. This huge new water source again brought riches to speculators increasing the boundaries and population of the City of Angels. As late as 1938, their relentless booster work was still in evidence, and was captured to the delight of Los Angeles citizens by journalist John Russell McCarthy in Los Angeles's popular *Pacific Saturday Night Magazine* in his tribute to William Mulholland:

> Some of us may doubt the necessity for building here, between the Pacific Ocean and Aimee Semple McPherson, a sanctuary for the outcasts of Pennsylvania, the sturdy sons and daughters of Iowa, and the trailer trash of the world. But Bill Mulholland never doubted. To his simple and rugged heart a great Los Angeles meant a big Los Angeles, and by the Eternal he would make a big Los Angeles possible.
>
> Even Owens Valley, for all its yells and dynamite, could not yield enough water to satisfy Bill Mulholland. Within ten years after Los Angeles had amazed the world (and the sands of the San Fernando Valley) by bringing the pure cold clear mountain water all those miles, he was up and at it again. Where to get more water? Well, there was nothing left but the Pacific Ocean and the Colorado River. The Pacific Ocean is useful in making bathing suits seem more-or-less nonexistent, but it makes bitter drinking. Mulholland simply went after the Colorado.

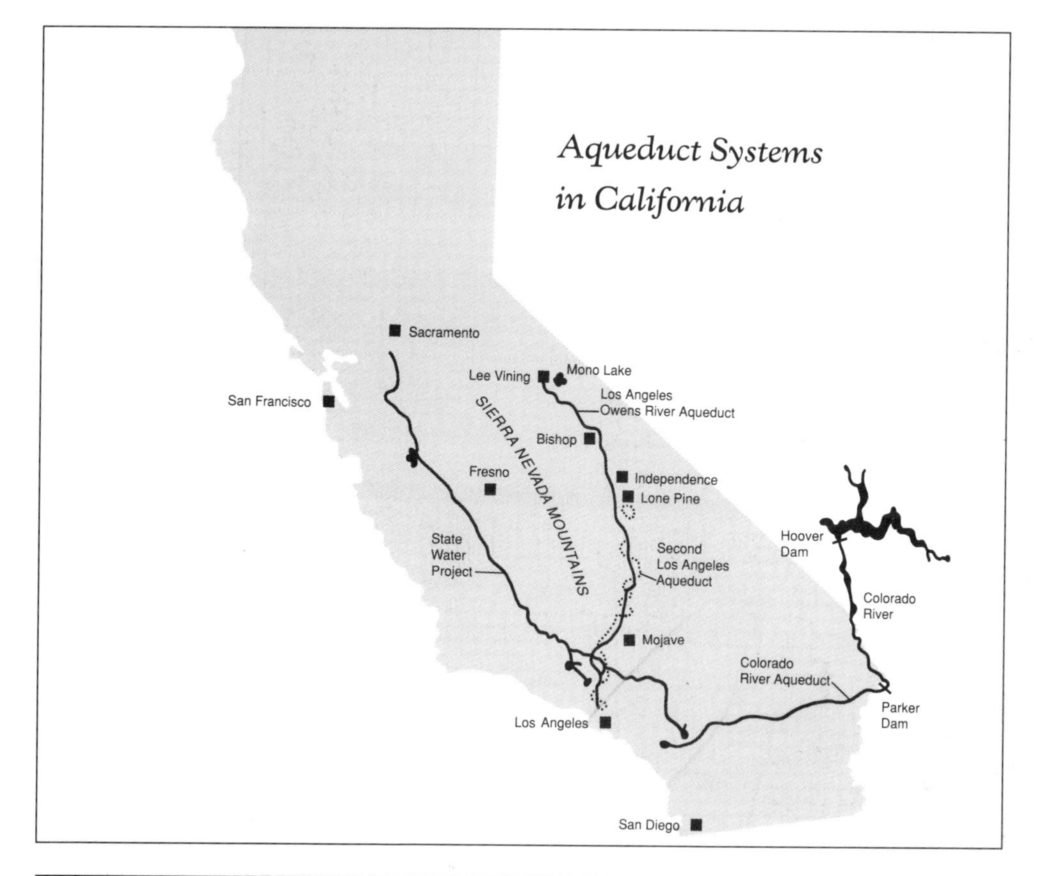

A contemporary map of the four major California aqueducts that provide Los Angeles's water: The Los Angeles Owens River Aqueduct, Second Los Angeles Aqueduct, The Colorado River Aqueduct, and The State Water Project.

William Mulholland, Superintendent of the Los Angeles Department of Water and Power, circa 1895. *(Los Angeles Department of Water and Power)*

The Owens River as seen from the northern juncture of the Los Angeles Aqueduct. Mulholland's challenge was to move this water 250 miles across mountains and desert to Los Angeles. Later, Mulholland was to pay a terrible price for his failure to acknowledge the rights of the Owens Valley.
(Los Angeles Department of Water and Power)

The trio of engineers who wrested water rights for the aqueduct *(from left to right):* Joseph B. Lippincott, Fred A. Eaton, and William Mulholland. This photograph appeared in the *Los Angeles Times* on August 6, 1906, one week after Lippincott's resignation from the United States Reclamation Service following charges of fraud and conflict of interest.
(Los Angeles Department of Water and Power)

The infamous Long Valley Cattle Ranch, which was to figure so importantly in both the achievement and tragedy of the Los Angeles Aqueduct. *(Los Angeles Department of Water and Power)*

Men at work at the south end of the Jawbone siphon, 120 miles north of Los Angeles. Water was moved through steel pressure siphons over mountains and canyons. The Jawbone siphon spanned 8,095 feet.
(Los Angeles Department of Water and Power)

An open-lined channel of the Los Angeles Aqueduct. The aqueduct consists of 250 miles of both open and covered channels, as well as siphons and conduits that carry water entirely by gravity from the slopes of the Sierra Nevada Mountains to Los Angeles.
(Los Angeles Department of Water and Power)

Construction camp at Big Creek, California. *(Los Angeles Department of Water and Power)*

Aqueduct workers pose atop an Electric Dipper Dredger in the spring of 1910. The dredgers were used to excavate canals for the Los Angeles Aqueduct. *(Los Angeles Department of Water and Power)*

Fifty-two-mule team hauls giant sections of steel siphon to the aqueduct site. Mule skinners were scarce who could handle such teams.
(Los Angeles Department of Water and Power)

Lucille Mulholland's secret wedding in 1919 to her second husband made the front pages of Los Angeles newspapers and infuriated her father. Mulholland immediately filed action against Lucille to gain guardianship of his infant granddaughter, the product of Lucille's first ill-fated marriage. A bitter custody battle ensued.
(Special Collections, University Research Library, UCLA)

Dr. Raymond G. Taylor, shown here behind the wheel of his Model D Franklin, was an integral part of the aqueduct story. Taylor said that his experiences as the aqueduct doctor, tending to the medical needs of 5,000 workers scattered along the construction route, afforded him the opportunity to participate in what he was sure would be one of the greatest engineering feats of the new century.
(Los Angeles Department of Water and Power)

An army of men had to be fed daily, a job which fell to food contractor Joseph "D. J." Desmond (not pictured), who was not always able to keep the men happy nor to account satisfactorily for his operation.
(Los Angeles Department of Water and Power)

The championship work crew that set a new record in hard-rock tunneling during the construction of the 26,870-foot-long Elizabeth tunnel, the longest in the Los Angeles Aqueduct system. The crews of John Gray and W. C. Aston feverishly competed to reach the tunnel's center. Gray is seen second from left. *(Los Angeles Department of Water and Power)*

Graves of construction workers. Five men died in the dangerous underground work. The markers are those of the men who died in the explosion at Clearwater tunnel on June 6, 1912.
(Los Angeles Department of Water and Power)

The long-awaited moment: Owens Valley water arrives through the aqueduct. Nearly 30,000 jubilant celebrants witnessed the event at the Grand Cascades in Owensmouth on November 5, 1913.
(Los Angeles Times)

The hero of the hour: William Mulholland displays his commemorative ribbon at the Grand Cascades celebration for the arrival of the water. "There it is, take it!" he yelled to the thousands gathered at the site as Owens Valley water made its appearance.
(Los Angeles Department of Water and Power)

William Mulholland at the peak of his career. Completion of the aqueduct had made him one of the most famous men in the west. The self-taught Irish immigrant was awarded civic and academic honors he never dreamed possible.
(Los Angeles Department of Water and Power)

The ill-fated St. Francis Dam soon after its completion in 1926. The dark triangular stain is from normal outflow.
(Los Angeles Department of Water and Power)

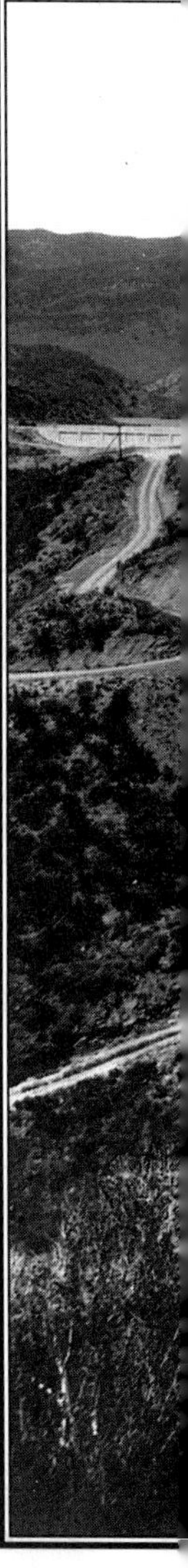

William Mulholland and city fathers with construction map of Mulholland Drive. Mulholland stands farthest right. *Los Angeles Times* publisher Harry Chandler is third from right; to Chandler's left stands General Moses Hazeltine Sherman, member of both the Los Angeles Board of Water Commissioners and the notorious San Fernando Valley land syndicate. Syndicate members made fortunes after the arrival of water to the valley and had ample reason to be grateful to Mulholland.
(*Los Angeles Times*)

View of the St. Francis Dam in progress, circa 1924, which Mulholland rushed into construction in order to impound a reserve of water for Los Angeles after saboteurs repeatedly dynamited the aqueduct. *(Los Angeles Department of Water and Power)*

William Mulholland and H. A. Van Norman view the wreckage of the dam early on the morning of March 13, 1928. The center section of the giant dam remained standing as would be the case if the two sides had been dynamited simultaneously, fueling speculation as to the cause of the dam's collapse.
(Los Angeles Times)

San Francisquito Canyon the morning after the collapse of the St. Francis Dam. Over 12 billion gallons of water roared through the sleeping valley wiping out everything in its path. Only the center section of the dam remained standing.
(Los Angeles Department of Water and Power)

The remains of Powerhouse Number Two below the St. Francis Dam following the disaster. The force of the millions of tons of water was so great that the solid masonry structure was completely obliterated, and nothing remained but the massive turbine housings.
(Los Angeles Department of Water and Power)

Jurors inspect the remaining center section of the St. Francis Dam before arriving at their decision as to what caused the dam's collapse and who was to be held responsible.
(*Los Angeles Times*)

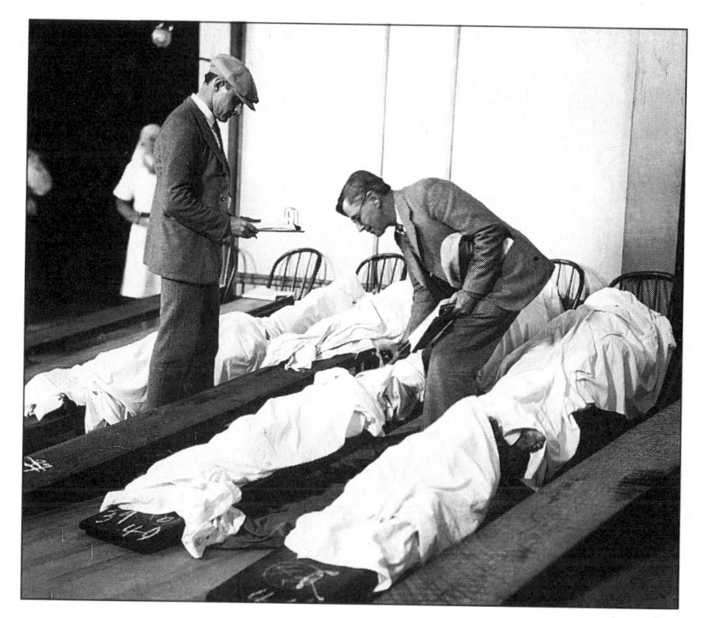

Undertakers catalogue the dead inside an emergency morgue in the town of Santa Paula. Two hundred bodies were recovered within twenty-four hours. As many as five hundred people were killed in the midnight flood. (*Los Angeles Times*)

William Mulholland on the stand, testifying at the coroner's inquest before Los Angeles coroner Frank Nance (*left*) and an inquest jury. Mulholland faced possible indictment for murder in the collapse of the St. Francis Dam. (*Los Angeles Times*)

William Mulholland examines evidence before the coroner's jury, March 20, 1928. (*Los Angeles Times*)

The Mulholland Dam and Reservoir deep in the Hollywood Hills. Mulholland's architectural and engineering showpiece, dedicated March 17, 1925, was renamed Lake Hollywood after Mulholland's fall from grace.
(Los Angeles Department of Water and Power)

Last photograph of William Mulholland taken at his St. Andrew's Place home in Los Angeles. Mulholland was in his seventies, ill and broken in spirit.
(Special Collections, University Research Library, UCLA)

11

UNBOWED

God is the judge:
He putteth down one,
and setteth up another.
Ps. 75:7

IN JULY, 1924, in the midst of the death threats and bombings, Mulholland received the painful news that John Gray, the man with whom he shared so much during their hard years of work on the aqueduct line, was dead. Gray had wandered into LaFayette Park near downtown Los Angeles, and shot himself in the head.

A simple note was found in Gray's pocket asking that in case of death the Ruppe Mortuary be notified. After Gray's body was taken to the coroner's office in the basement of the Hall of Justice, it was moved to the mortuary by the police.

The owner of the mortuary, L. E. Ruppe, stated to the officers that he had known Gray for many years following his work on the aqueduct. Ruppe told them Gray's wife had died a year earlier, and Gray had recently undergone surgery and had been in ill health for months. Ruppe was unaware that Gray had any children, but other sources believed that his son, Louis, had died in a construction accident a few years after the completion of the aqueduct. According to Gray's wishes there was no funeral and Ruppe took it upon himself to bury him in a simple grave.

Mulholland wrung his handkerchief when he read in the *Los Angeles Times* of Gray's suicide. He found it hard to imagine the indomitable Gray, the hard-working foreman who shoveled dirt and granite alongside his own men to keep their spirits from flagging during the labor strikes, would end his life defeated and alone. Overcome with grief, Mulholland could only wish that he had been at Gray's lonely grave site to say some final words as he had for those men who had died in his service on the aqueduct line.

A few months earlier, Mulholland had broken ground for a new holding dam at San Francisquito Canyon. During the next two years he would spend nearly every day supervising its construction. Driven by his chauffeur or by his son Thomas, he left his St. Andrew's Place home each morning at dawn, traveling forty-five miles northeast from Los Angeles to the Santa Clara Valley.

The dynamite attacks upon the aqueduct that summer had spurred Mulholland's rush to secure a giant storage reservoir hundreds of miles south from the battle zone in the Owens Valley. Mulholland had selected the canyon for a number of reasons: A more optimum site, located in Big Tijunga, was too expensive after owners had raised land prices to exorbitant levels. The canyon was located next to Powerhouse Number One, the main power facility for the San Fernando Valley, making it cheaper for the reservoir to generate hydroelectric power. There had always been concern for the aqueduct at the point where it transverses the San Andreas fault in the five-mile Elizabeth Tunnel, and need for an emergency storage facility below the fault line was undoubtedly one of Mulholland's considerations in selecting the San Francisquito Canyon site, but the unpredictable events in the Owens Valley that threatened the safety of the city's water supply was his primary concern.

"When these facilities have been put into commission," Mulholland declared, "the whole city will have been safe-guarded by storage near the south end of the aqueduct with a full year's supply of domestic water." The San Francisquito Canyon site offered another solution to Mulholland's predicament. Located within yards of the aqueduct, the canyon was large enough to house the size dam needed to solve Mulholland's water flow control problem, making construction of a dam at Eaton's Long Valley unnecessary.

Traveling to the canyon on the day of the news of John Gray's death, Mulholland's thoughts flashed to 1908. In that year, a large camp of aqueduct men laboriously at work on the Elizabeth Tunnel, including John Gray, had been housed on the floor of the future reservoir, and it was then that the idea of utilizing the site as a future storage facility first occurred to him. Mulholland had sunk shafts and tunnels deep into the red conglomerate surface to test its characteristics, and was convinced that the site was sound.

Mulholland knew the sufferings of others, and felt a sense of deep, embittered sorrow at the reasons behind the death of John Gray. His face lined with pain and eyes dark, Mulholland arrived at the site and focused his thoughts on the engineering problems of the St. Francis Dam.

MULHOLLAND WAS NOW past retirement age, but stood at the pinnacle of his career, "basking in the affection of his fellow Angelenos." His word was now gospel, and as his longtime secretary, Burt Heinley, wrote, recalling the criticism concerning the concrete used to build the aqueduct, "If Bill Mulholland should say that he is lining the aqueduct with green cheese because green cheese is better than concrete, this town would not only believe the guff but take oath that it was so." Since the turn of the century, Los Angeles had grown five times its size, and the city surpassed San Francisco as the preeminent port on the West Coast. The aqueduct had established the San Fernando Valley as an unprecedented agricultural region firmly establishing Los Angeles's new posture as the number-one agricultural county in the nation.

But the forces that had plagued Mulholland since 1920 had come to a head in 1924, ultimately forcing him to do the one thing he always resisted—surrender to political pressure from city bureaucrats and wealthy business interests. The dual threat of continuing drought and the continued war with the Owens Valley diehards posed a mortal danger to the city's water supply and its economic prosperity, and city leaders urged Mulholland to immediately secure substitute reservoir facilities. Even when bombings of the aqueduct ceased, Mulholland's fears of the ever-present threat of drought made it necessary to construct a buffer water supply system to safeguard the city's resources.

Without Eaton's Long Valley or another northern reservoir, flow from the aqueduct into Los Angeles could not be adequately regulated. At Haiwee Reservoir, as the system was presently designed, the aqueduct water could not be sufficiently controlled, and huge quantities of water overfilled the San Fernando Reservoir whenever the snow pack was bountiful in the Sierra Nevada, forcing engineers to dump thousands of gallons each day into the Pacific Ocean to ease the immense pressure on the dam wall—a practice Mulholland deplored.

Mulholland preferred pumping water back into the large natural underground lakes of the local aquifer, a concept since embraced by hydrologists. But civic and business leaders, including the office of the mayor and the Board of Control pressured Mulholland and the Board of Water and Power Commissioners to construct a string of reservoirs throughout southern California. In the event the aqueduct was damaged or a severe drought persisted, the city of Los Angeles would have a water supply sufficient for at least two years. Immense economic and political interests were at stake, and the pressure placed on Mulholland to act quickly was enormous.

In 1923, Mulholland began construction of the Stone Canyon, Encino, Silver Lake, and Mulholland Dam reservoirs. In August, 1924, concrete was poured at San Francisquito Canyon dam, located midway between Bouquet Canyon and Elizabeth Lake. The dam was named the St. Francis, and was intended to be the largest arch support dam in the world, standing nearly 200 feet tall, 700 feet long, 176 feet thick at its base, with a reservoir capacity of 38,168 acre-feet, covering 600 acres. When it was finished it had cost $1.3 million to build, and became the second-largest reservoir in Mulholland's system, holding more than 12 billion gallons of water, second only to that at Haiwee.

However, Mulholland's hasty construction of the secondary holding sites did not deter the irate Owens Valley ranchers, who continued their campaign of sabotage, keeping Los Angeles and Owens Valley in intermittent states of siege.

In the hottest month of the summer of 1924, there were six dynamitings, and at dawn on Sunday, November 16, 1924, led by Mark Watterson, seventy armed men from Bishop drove a caravan of dust-covered Model-T Fords to the Alabama Gate spillway, one of the strategic water diversion controls of the Owens River aqueduct, just north of Lone Pine, and seized control of the gate house.

The insurgents opened the hydraulic gates and hundreds of thousands of gallons of water crashed down the open spillway onto the desert floor. This relatively simple act of terrorism completely suspended water service to the entire city of Los Angeles.

The "rebellion," as it was deemed by the press, overnight became a publicity bonanza, creating a media event of such proportions that newspapers in Paris reported it. Suddenly the eyes of the world were riveted on "California's Little Civil War." The armed seizure of a municipal water system violated local and state law, but the "felonious" scene at the spillway was quickly transformed by Owens Valley citizenry into a celebration of civic solidarity. By noon the following day, seven hundred men, women, and children had joined the insurgents in a massive impromptu picnic at the spillway site. Western movie star Tom Mix, filming nearby in Bishop, brought his camera crew and a mariachi band to join in the festivities. Hoping to win public attention and sympathy for their plight and force the state government to intervene, the insurgents began camping out at the Alabama Gate with their families and townspeople. In the four days and nights to come, the picnic turned into a huge cookout that captured the imagination of headline writers everywhere.

The Los Angeles County Sheriff's Department and Mayor George

Cryer of Los Angeles pleaded with California Governor Friend Richardson to dispatch the militia to disperse the ranchers and regain control of the gate house. Fearing potential bloodshed and the negative publicity surrounding the deployment of state troopers, Richardson refused. When an Inyo County Superior Court judge was pressured by Los Angeles authorities to issue the insurgents' arrest warrants, he readily disqualified himself from the case, and the Inyo County sheriff refused to interrupt the proceedings, stating that he was a "friend and sympathizer."

No arrests were made, and the jubilant insurgents stood in all-night vigils, directing searchlights into the sky as local merchants delivered freshly slaughtered beef, pigs, and chickens, loaves of baked bread, milk, and hard liquor. The celebration continued, and eventually someone posted a sign in the center of a deserted Bishop exclaiming, "If I'm not on the job, you can find me at the Aqueduct."

The Owens faction, sensing victory at last, were now drawn together in sentiment and common purpose more than ever before. W. A. Chalfant, editor of the *Inyo County Register*, hailed the insurrection as an "American community . . . driven to defense of its rights."

Behind the festive scenes reported in headlines to a sympathetic world, inside the Department of Public Service there was a sense of cold fear and political uncertainty. When Mulholland learned that the man recruited to open the hydraulic gates had been an engineer who had worked on the aqueduct, he was furious at the betrayal. He immediately ordered the length of the aqueduct guarded twenty-four hours a day by private police, paid for by the city of Los Angeles.

"Interference with a public utility," Mulholland angrily declared to the press, "is a serious matter—of greater import than interference with the United States mail. . . ." Mulholland met in emergency sessions with the Board of Public Service Commissioners, the office of the mayor, and representatives from the governor's office, and pronounced that lawless elements of a "radical fringe" had taken over and must be stopped at all costs.

Expecting public concurrence with his righteous rage, Mulholland was stunned to see that by the fourth day of the Alabama Gate seizure, newspapers shifted to recounting sympathetic human interest details of the rebellion by an "oppressed pioneer community." From the *Los Angeles Daily News*:

> An orchestra with several khaki-clad girls dragging drums and musical instruments with them, climbing up the slope a chord or two is sounded and 350 lusty throats joining in the singing of "Onward, Christian Soldiers," the song sweeping across the barren Owens Valley and re-echoing against the tinted hills of the east.
>
> Mid-day passed, the sun dips behind the suncrest of the high Sierras. Long shadows creep across the valley as chilly winds, kissed by the snows, slip down from the mountain heights. Dusk settled and the determined ranchers remained at their posts. Bonfires began to grow on the hill sites, and a welcome aroma of hot coffee drifts across mesquite stretches and the overall ranchers remain on into the night.
>
> Darkness descends and the women folks, wearied by their day's toll feeding their men—the guardian of their homes—join their husbands about the fires. Some are nestling their babies, others are "feeding" a phonograph with operatic selections.
>
> And the men of the Alabama toll gate remain on. Talking quietly in groups and never asking—how long?
>
> Who are these people who have hurled the boldest defiance at the officials of Los Angeles? They are the women of Bishop, wives of ranchers, Bishop's leading business and professional men who have taken it upon themselves the big job of feeding scores of Owens Valley men who have opened the flood gate. The city can afford to be liberal in its settlement with these pioneers whose work of half a century it will undo.

When still other area newspapers reported stories sympathetic to the rebels, Mulholland became incensed, blasting the newspaper accounts as "pure hogwash." The facts, however, spoke for themselves. The resistance movement was based on one simple premise: The Owens people could tolerate Los Angeles's compulsion to look to the valley to supply its water, but they could not justify their own demise when it could be avoided.

The editorial page of the *Los Angeles Times* further summed up a general feeling among Los Angeles citizens about the unrest in Inyo County:

> It is to be remembered that these farmers are not anarchists nor bomb-throwers, but, in the main, honest, earnest, hardworking American citizens who look upon Los Angeles as an octopus about to strangle out their lives. They have put themselves hopelessly in the wrong by taking the law into their own hands but that it is not to say that there has not been a measure of justice on their side of the argument. . . . There must be no civil war in California.

The images of a tiny community united in defense against a Goliath oppressor were fed into the consciousness of the people of Los Angeles and the world. But for William Mulholland, the events at the Alabama Gate were nothing more than a lawless mob seizing government property for publicity, and he was fiercely determined to do everything within his power to secure order and cease any interference in the operation of his city's aqueduct.

12
RETRIBUTION

Consider mine enemies,
for they are many.
Ps. 25:19

THE TROUBLES IN OWENS VALLEY momentarily seemed but a distant popping of a harmless holiday firecracker. Though shaken by the recent uprisings, Mulholland looked forward to the approaching New Year's holiday with noticeable good cheer. In a Mulholland-sized Christmas present to the much-honored chief himself, more than ten thousand people gathered on December 28, 1924 to see Mulholland dedicate the Mulholland Skyline—a million-dollar majestic highway riding the twenty-two-mile crest of the Santa Monica mountains from Hollywood to the Pacific Ocean.

Befittingly, the occasion was epic in scale. Two brilliant, silver-caparisoned Spanish caballeros mounted on horseback stood guard while Los Angeles City Police Chief Roderick Heath, as brilliantly caparisoned in his sparkling blue uniform, handed William Mulholland a large golden key to unlock the exotic flower-bedecked portal erected in the middle of the highway. Mulholland, in a firm movement, smashed a bottle of Los Angeles Aqueduct water over the key and then inserted it into the gold lock. The chain of flowers attached to the sides of the portal and stretching to the shoulders of the highway fluttered apart, and the petals floated into the air like multicolored doves. Thousands of men, women, and children broke into a deafening cheer as the portal doors opened and the long procession of cars began to pass through.

The most important city officials, including the mayor, and members of the omniscient Board of Control led the procession in their gleaming black limousines. As they passed heading east to Hollywood, a squadron of low-flying airplanes dipped their wings in salute and then roared away. The procession traveled the length of the new highway, the "rim of the world," as the papers grandly called it, encircled by verdant foliage, recently planted, panoramic views of the San Fernando Valley to the north, and twisting curves to the south—it was a breathtaking stretch of wonderland.

Reaching the end of the highway—a bluff overlooking the panoramic views of Hollywood and the Pacific Ocean beyond—the caravan entered Laurel Canyon and wove down its twisting lanes into Hollywood. There they were greeted by celebrations complete with a rodeo, and the squadron of "birdmen" stunt flyers dipping and careening in the blue, smogless Los Angeles sky. Further tribute to the magnificent highway was followed by a giant parade down Hollywood Boulevard from Vine Street to Highland Avenue. Mulholland, seated imperially in an open limousine was, of course, the Grand Marshal.

Following respectfully behind the man of the hour were civic leaders, members of the Board of Public Works and the man who actually built the highway, engineer DeWitt Reaburn. Befitting an eastern potentate, the parade included a squad of mounted police, followed by uniformed officers and bugler; marching behind them was the Third Coast Artillery Regiment, with a mobile searchlight, a field cannon drawn by military tractor, and a motorized antiaircraft gun. Behind them came a full naval band, five thousand sailors, two hundred marines, and bringing up the rear, bagpipes, mariachis, and the American Legion Band.

Like a huge glorious caterpillar, the parade turned north on Highland Avenue and ended at the beautiful, two-year-old Hollywood Bowl, where Mulholland gave a charismatic speech to the cheers and shouts of the overflowing outdoor audience. Toasts and tributes were followed by vaudeville acts, speeches, fireworks, and more speeches.

Characteristically, Mulholland, in his address to the throng, disclaimed any personal credit for building the highway, giving credit to DeWitt Reaburn standing in the crowd of dignitaries behind him. By now at this stage of his long and prominent career, tribute to his ears was an old song. But rising to the occasion, the Chief modestly admitted that he did conceive the idea of the highway thirty or forty years earlier.

More than ten thousand people flocked to the Hollywood Bowl to see and hear the great man for whom "one of the most scenic wonders of the southland" was named. Mayor George Cryer pronounced that the opening of the highway was the crowning event in engineer Mulholland's career. The Commander of the United States Pacific Fleet, Admiral George Robinson, declared that today's engineers of the great United States of America were the real "conquerors of today," comparing them to the "conquerors of yesterday," the ancient Roman armies which drove their slaves into the building of the Roman highways. Mulholland was heralded as the consummate American dreamer, who, when most people thought of Los Angeles in terms of only a hundred thousand inhabitants, visualized it as a grand metropolis for two million souls and more.

The opening of the highway was one of the largest celebrations in the history of Los Angeles. The gala continued non-stop until midnight, whereupon Mulholland returned home, no doubt intoxicated by the heady acclaim as much as he was by the liberal toasting of alcohol. By morning, his attentions were again focused on the problems at Owens Valley.

IN OWENS VALLEY, ranchers continued to institute a host of legal reforms to secure a continued indigenous water supply for the valley's domestic and agricultural needs. Mulholland's defense against this strategy, according to author John Walton, was "to humor the resistance movement with protracted negotiations over its demands, particularly the question of dam and storage reservoirs while simultaneously undermining its solidarity with individual property purchases." Unable to compromise with Eaton over his million-dollar demand, Mulholland apparently, as critics pointed out, "preferred the slow economic death of the Owens Valley to Eaton's profit in Long Valley at city expense."

The insurgents achieved a national audience with their takeover at the Alabama spillway. Owens Valley newspaperman W. A. Chalfant had written to Mary Austin that the "incident at the spillway became . . . a 'shot heard 'round the world,'" and the Owens Valley rebels hoped that the state of California would be forced to intervene and disperse the militia, thereby guaranteeing even more attention to their plight. But, fearing widespread violence, Governor Richardson shrewdly refused to take the bait, and when he did not send in state troopers, he upset the rebels' plans, and by the fifth day the popular citizens' takeover ended without violence. Talks began a week later, but broke down after two days. In the end, after innumerable negotiations, nothing was settled, and the city continued its piecemeal purchases of valley land connected to the Owens River.

Anticipating even more violence, Mulholland intensified efforts in the Santa Clara Valley to complete the St. Francis Dam as quickly as possible. By spring 1925, Mulholland Dam in Hollywood, one of thirteen proposed reservoirs in the buffer system, was finished. The $1.25 million dam was formally dedicated on March 18 in a simple yet impressive ceremony. Several hundred prominent citizens including Mayor George Cryer, Harvey Van Norman, W. B. Mathews, J. B. Lippincott and two members of the Board of Control were present.

The ninety-acre Lake Hollywood impounded behind Mulholland Dam today stands beneath the huge white letters of the Hollywood sign. The dam that Mulholland designed endures as an exquisite creation, an example of municipal Mission-style architecture. "From the north, the parapeted walkway, the balustrades, and single flanking tower give the dam an appearance of a castle wall and turn the lake into a broad moat. A row of concrete bear heads extends over the arches on the south side of the dam in homage to the state's flag . . . a dam with panache, a memento of an era when the Department of Water and Power's resources were allocated for the beautiful as well as the useful."

"When I built this, " Mulholland said in his dedication speech, "I wanted something ornamental and architectural as well as useful. The structure is itself the best tribute."

Mayor Cryer lauded the beauty of the work in dedicating the structure to the man who conceived the aqueduct: "For forty years Mr. Mulholland has been the guiding genius of the city's water organization, giving all that is in him to upbuilding of the municipality and in the service of the people. We are here today to dedicate this, the latest monument to his skill, his knowledge and his unquestioned ability. It probably will last for centuries. So I say: 'Hail to William Mulholland, long may he live and prosper.'" Following the dedication of the dam bearing his name and the unveiling of the bronze tablet at the side of the structure's tower, Mayor Cryer opened the gates that permitted the water to flow into the city's water distribution system.

"It hasn't rained for three years," Mulholland announced in his acceptance speech. "Yet you have impounded by this dam water which has been brought to this spot for a distance of 238 miles over mountain and desert, over vast, uninviting areas. This water comes from the San Fernando Valley where it has already been used once to grow crops that add materially to the prosperity of our city and southern California." Mulholland's remarks were followed by a standing ovation. "This entire area," read the official program, "is a natural enclave of serenity and tranquility hidden in the heart of the city."

One year later, with the deteriorating situation on his hands in the

Owens Valley, and the looming threat of more bombings, Mulholland did not indulge in any publicized ground breaking of the St. Francis Dam when it was completed. Instead Mulholland quietly christened the enormous project with a bottle of champagne, surrounded only by a handful of invited guests and engineers from the Department of Water and Power. By winter, Mulholland commenced the slow, tedious process of filling the structure with Owens Valley water from the Los Angeles Aqueduct, completing the string of emergency reservoirs demanded by city boosters.

Mullholland hoped that the St. Francis Dam would close the ugly chapter on the controversial Long Valley Dam, and solve the vexing problem of an adequate storage site which had haunted his reputation and the aqueduct's success. It was another crowning achievement in Mulholland's lengthy, illustrious engineering career.

IN JANUARY 1926, Fred Eaton traveled by train from his Long Valley ranch to Los Angeles to conduct business and visit two of his children. While he was away, two board directors of Eaton's Land and Cattle Corporation, in collusion with other officers of the firm, took out a $200,000 mortgage against his ranch in the company's name.

Mark and Wilfred Watterson knowingly issued the loan to Eaton's associates without Eaton's legal approval. Once the cash had been distributed, Eaton's land was encumbered by a $200,000 note owned by the Wattersons' bank. Shortly thereafter, the Wattersons sold the note to the Pacific Southwest Trust and Savings Bank of Los Angeles.

Unaware of the fraudulent transaction, Eaton remained in Los Angeles for two weeks doing business related to his cattle interests, visiting the dentist and ordering new clothes. He made no attempt to see William Mulholland, although as former city mayor, he entertained and dined with several prominent city leaders. Untypically, Eaton did not promote sale of his Long Valley dam site on this particular visit, but his fixation on the goal of obtaining his "just price" for the property persisted. At first, Eaton fixed a price of $900,000 for the land, but city officials appraised the value at no more than $255,000. After Mulholland's continued refusal to meet this price, anticipating a rise of value due to the difficulties in the Owens Valley and the eventual return of seasonal drought to the Los Angeles basin, Eaton raised the ante to $1.5 million, then doubled the amount to $3 million.

By now Eaton could well afford to play a game of bluff. He owned more than 12,000 acres, 5,400 head of cattle, and a sprawling white-walled hacienda ranch house surrounded with bougainvillea, fig trees, and fruit orchards. His cattle empire on the banks of the Owens River was thriving, and Eaton was dubbed by Hearst's *Examiner* as the "Cattle King of California." Eaton's son, Harold, in a role much like Mulholland's dutiful son Perry, was manager of the Eaton Land and Cattle Company, and father and son devoted themselves to working its rich pasture lands and bountiful watersheds.

Despite these comfortable assets, Eaton still envisioned more money for himself, and speculated that in the future, downstream cattle ranchers would soon be forced out of business by the rapidly depleting indigenous water supply in the valley, and would have no choice but to sell their cattle to him at drastically reduced prices, further bolstering his considerable wealth. Eaton was convinced the longer he maintained control of Long Valley, the longer he held the golden key of opportunity. With Long Valley, Eaton was sure he held the city of Los Angeles captive for a big future pay-off, and believed he would become the sole surviving rancher in the valley while the others died a slow economic death.

As a result, seventy-one-year-old Eaton was in no real hurry to cash in. He was a millionaire on paper, and biding his time, knew Los Angeles would have to pay his price. Returning to his ranch, Eaton continued to relish the activities of his cattle empire with his son, maintaining his dream of further riches and bringing Mulholland, his once pupil and now nemesis, to his knees. It was a coveted delusion that would last for ten years.

Harold Eaton learned about the $200,000 note taken in secret by other corporate officers of the Eaton Land and Cattle Company. When Harold told him about the swindle, his distraught father suffered a stroke that rendered him unconscious and paralyzed. The family remained at his bedside in a twenty-four-hour vigil fearing he would die.

Knowing the financial crisis that would soon come crashing down upon the Eaton family, Harold authorized his father's longtime associates George Khurts and J. T. Fishburn to negotiate on behalf of the stricken Eaton a quick sale of the Long Valley dam site. Harold then immediately wired Mulholland at the Department of Water and Power informing him that Eaton was critically ill.

The news of Eaton's stroke left Mulholland deeply shaken. As a gesture to his ailing former friend, Mulholland, in a softening of his earlier posture, instructed Van Norman and W. B. Mathews to try to see to it that the city now make a "fair" offer on Eaton's Long Valley. Van Norman and Mathews immediately began negotiations with Eaton's representatives. By March 11, 1926, it appeared as though Eaton's people would authorize a price of $66 per acre, roughly $792,000, but Van Norman knew that the commissioners and the city council would never accept a price in excess of $40 per acre, or $480,000.

On March 12, Eaton suffered a second debilitating stroke. Hoping that resolution of the problems at Long Valley would encourage his father's recovery, Harold turned to others and began heated and frantic efforts to close some kind of deal. Wishing to influence Mulholland, Harold wired W. B. Mathews in Washington, D.C., about his father's dire condition and pleaded for a swift resolution of the problem. Mathews quickly wired to Harold that he was available to resolve the matter and conveyed "his affectionate regard and earnest hope for an early recovery."

On March 26, Harold Eaton reversed tactics, feistily demanding that the city accept the offer at $66 per acre now, or he would tie the property up for five years by selling options on parcels to a land syndicate of Owens Valley ranchers, which had been formed by the Watterson brothers. Van Norman, sympathetic to Eaton's plight, wired W. B. Mathews that he frankly believed that Harold's threat was another bluff. But in no event, he said, could the Department of Water and Power justify Eaton's price. Earlier, under the supervision of Van Norman, three teams of appraisers had evaluated the property and found its value to be no more than $40 an acre.

By early January 1927, Eaton was slowly recovering and had regained partial physical mobility, but no deal had yet been struck. By mid-January, Harold reiterated to Van Norman his father's price of $800,000, as opposed to his original $1 million. As Los Angeles officials continued to mull over the offer, Fred Eaton suddenly raised the ante to $1 million once again.

To Eaton's disadvantage, there had been no further dynamitings in the Owens Valley since the final blasts at the Alabama Gate on May 12, 1926. By now, Mulholland had completed the string of reservoirs to strengthen the city's emergency water supply, and the city was no longer as desperate to obtain the Long Valley dam site.

BUT AFTER ONE YEAR OF CALM, in May 1927 violence erupted again against the aqueduct; suddenly, Eaton's ambitions were closer to becoming reality. Four masked men kidnapped two guards at the No Name Siphon, ten miles south of Little Lake, and dynamited 450 feet of concrete aqueduct pipeline. With precious water gushing from the immense rupture onto the desert floor engineer Clark Keely arrived on the scene and described it as a "torrent of water . . . a young Mississippi flowing out over the desert."

Mulholland ordered machine guns posted at every conduit and horseback patrols to guard the length of the pipeline at hourly intervals; Harvey Van Norman was dispatched to direct operations. Floodlights were installed throughout the system to illuminate trespassers, but bullets from out of the darkness of night shot the lenses out. A second explosion blew up the intake at the power plant at Big Pine Creek. Then at Cottonwood, twelve miles below Lone Pine, a concrete aqueduct wall was dynamited. Mulholland's aqueduct, now listing like a huge tanker under siege by submarines, was severely damaged and the lifeline of the Los Angeles water system was incapacitated.

Mulholland responded with a vengeance. He ordered six guards armed with Winchester rifles and tommy guns dispatched to No Name Canyon with orders to shoot and kill any suspect, no questions asked. The city offered a ten-thousand-dollar reward for information leading to conviction of the insurgents, and Mulholland called upon the governor to send in state and federal troops. The *Los Angeles Times* reported that six hundred specially recruited police were "concentrated and ready to do battle." Within five days, Mulholland sent another hundred guards to patrol the entire length of the 233-mile aqueduct, and scores of uniformed patrols roamed up and down the valley at all hours with searchlights. Owens Valley was now in "a state of virtual warfare." Mulholland believed that nothing short of an armed occupation by city police would stop the violence.

As Mulholland's armed patrols gathered in one place, sabotage occurred in another, and the bombers' ingenuity turned the patrols into a vicious game of hide-and-seek. The aqueduct was hit with four more blasts. Two miles south of Lone Pine, at Tuttle Creek, a sixteen-foot hole was blasted open. Two bomb explosions occurred at the Alabama Hills, and then, on July 16, a forty-foot section of the pipeline was blown to pieces at Tebo Gates. Investigators discovered even more bombs along the line which luckily had failed to explode. Los Angeles–owned water wells in the Owens Valley were incapacitated, aqueduct phone lines were cut, and police search lights dismantled.

Finally, in view of the destruction, newly elected California Governor Clement Young asked President Coolidge to intervene, and the White House dispatched uniformed federal agents and Pinkerton men to patrol the aqueduct. In retaliation, bombers struck the Big Ditch section of the aqueduct near Lone Pine. Blasting the aqueduct—or "Shooting the Duck" as the bombers proudly called it—became the obsession of the insurgents throughout the long, violent, hot summer of 1927.

Much to the fears of Los Angeles city officials, the violence appeared to be spreading to Los Angeles itself. An anonymous phone call was received by the Los Angeles Sheriff's Department reporting that a carload of armed men were on their way from Inyo County to dynamite the St. Francis Dam. Due to the dynamiting in the Owens Valley, the St. Francis Reservoir was now the sole source of water from the north, and a threat of sabotage to it was enough to shake even Mulholland's steel nerves. At his insistence, dozens of officers descended on the high concrete ramparts of the St. Francis. Apparently tipped off to the police response, the bombers never materialized.

After federal troops arrived on the scene, Mulholland's anger turned into grave concern over the future protection of Los Angeles's fragile buffer water supply. By this point, the St. Francis Reservoir was almost filled to capacity, and Mulholland ordered heavy water withdrawals from the reservoir to replace the lost flow into Los Angeles from the damaged aqueduct. Mulholland's only source of consolation during this unpredictable period was that in one short year, the St. Francis Dam had already justified its existence by feeding a parched city which was now threatened with outright water famine. As the violence intensified and the citizens of Los Angeles realized their well-being was in the hands of "a group of angry valley terrorists," public sympathy for the ranchers quickly dissipated.

But the super-charged atmosphere in the valley remained so heated that it was ripe for bloody confrontation, and Mulholland, pacing in anger in Los Angeles, was fiercely determined to end the war once and for all. He now directed his attention to attacking what he believed to be the root reason underlying the unrest in the valley—the economic power and resulting control the Wattersons maintained over Owens Valley residents.

According to historian William L. Kahrl, Mulholland obtained detailed financial statements on the Wattersons' operations which suggested that some bank funds had been diverted illegally to other Watterson enterprises. On August 2, Mulholland took the evidence to the state corporations commissioner, who dispatched a state bank investigator to the Owens Valley. Three days later, while an audit ensued, the Watterson banks closed and on August 10, the brothers were jailed on charges of embezzlement.

In the subsequent trial, Wilfred and Mark Watterson admitted they had diverted more than $2.3 million of the ranchers' savings into their own companies, with an explanation that they had acted only to "preserve valley industries in the face of the city's onslaught." The brothers were convicted on thirty-six counts of embezzlement and grand theft, and were sentenced to ten years in prison.

There was scarcely a rancher or merchant anywhere in the valley who did not have a mortgage with the Watterson banks. Hundreds of Inyo County residents, who had placed their life savings with the trusted Watterson banks, were ruined, their savings lost, and their property forfeited, and the valley economy virtually collapsed overnight. With the banks' demise, the resistance movement was broken, and the violence against the aqueduct and Mulholland abruptly ended.

Fred Eaton did not escape the fate that he had once wished upon his valley neighbors. When the Watterson banks collapsed, the $200,000 note secured by Eaton's Long Valley property went with them. The Los Angeles bank holding the note on Eaton's land promptly initiated foreclosure proceedings. Unable to come up with the $200,000 cash required to save his ranch, Fred Eaton suffered a third stroke. He and his family now lived in constant fear that they would be forcibly removed from their home by the Owens Valley County Sheriff and stripped of the possessions of a lifetime.

ONCE AGAIN, an uninterrupted flow of water coursed from Owens Valley to Los Angeles. The water was secure but it was becoming increasingly apparent that it was insufficient to meet the city's demands. Mulholland now focused his thoughts on another struggle —his continuing fight for the Colorado River project. After three years of intensive lobbying in Sacramento, Mulholland had finally obtained the California state legislature's approval of a bill creating the Metropolitan Water District, thereby laying the necessary groundwork for the creation of the Colorado River Aqueduct.

On New Year's Eve, 1927, Mulholland, Van Norman, and Mathews set out in a first-class sleeping cabin on the Southern Pacific for an eight-day trip to Washington, D.C., to lay their request before the U.S. Senate. Before departing from Union Station, Mulholland declared with finality on the steps of his rail car to a throng of reporters: "Members of Congress and people of the Southwest must realize that this section of the country is face-to-face with a desperate and dangerous situation and only the federal government can aid us."

This trip was one of dozens Mulholland had made to Washington with W. B. Mathews to testify on behalf of the Swing-Johnson bill, the final shot in the six-year legislative battle for Boulder Dam. Residing at the elite, staid Cosmos Men's Club, for months Mulholland and Mathews lobbied, entertained, and testified in endless hearings on the Hill as the bill wound its way tortuously through Congress.

Upon return to Los Angeles from one of his lobbying junkets, Mulholland was surprised to find the now seventy-three-year-old Fred Eaton calling upon him at his offices at the Department of Water and Power. Eaton had recovered sufficiently from his last stroke to be treated by doctors in Los Angeles. Mulholland was saddened almost to tears to see the once handsome, vibrant comrade of earlier days demonstrating the obvious signs of partial paralysis—the slurring of speech, facial distortion, and incoherent thoughts.

Eaton again reiterated his demands for Long Valley. Pridefully he kept the impending foreclosure of the property to himself, refusing to sell it to Mulholland for less than a million dollars. Mulholland had no choice but to deny Eaton's demand. In spite of his illness, Eaton issued forth a rain of invective. In a replay witnessed many times before by the hapless Mulholland, the infuriated Eaton stormed out of Mulholland's office.

A few weeks later, Mulholland received an urgent telephone call from the wife of one of Eaton's sons, Burdick, who lived in Los Angeles, requesting that Mulholland come to the house to visit Fred as soon as possible. In a letter Mulholland wrote to W. B. Mathews still lobbying feverishly in Washington, D.C., Mulholland described his visit to Eaton:

> Burdick's wife called me last Friday to come and see Fred. I was somewhat alarmed by the urgency of her call and I responded at once, but found when I got there that he was merely in a talkative mood.... He wanted to prod me up about his affairs. He doesn't seem to me to be ailing anymore than he did a couple of weeks ago when he called here at my office, but seems to realize that the time is getting short to urge his case in the land matter on the Board....
>
> As you know, I am keeping clear of it altogether myself, except when urged to talk about it, as I was on my visit to Eaton, but I avoided discussion of the matter as much as I could. I told him he would have to take it up with the proper committee. The Board has not approached the subject in my presence at any time within the last three or four months, or at any time since you left here. My purpose in writing you this, is, as I stated, so as to let you know how things stand as far as anything we have done here is concerned....

Eaton's "desperate" call to Mulholland can be taken as more than just erratic behavior of a stroke victim. Even though he had stormed out of Mulholland's office angry and defiant, the underlying reason for the visit was perhaps an unexpressed cry for long lost companionship. Besieged by impending financial doom and death, he could find comfort in the company of an old friend-turned-enemy.

As Mulholland implied in his letter to Mathews, the matter of Long Valley was no longer completely in his hands. Mulholland's vote either way would now not make it possible for Eaton to get his price. The water board had the final say and Eaton, volatile and erratic in his dealings with them over long years of stormy negotiations, had sealed his fate with some of the members.

Sympathetic to Eaton's dire plight, Mulholland dispatched Van Norman to visit the ailing Fred Eaton and try to come to some kind of an arrangement. "Going to see Fred and see if anything can be done," Van Norman telegrammed Mathews. "I think the matter can not be handled until the situation becomes so desperate for Fred that he will agree to let some of his friends work out a deal without his interference." In deference to Eaton's former position as mayor of Los Angeles and superintendent of the Los Angeles City Water Company, his ailing health, and Mulholland's concern, Van Norman continued with: "I hope it can be done with sufficient clear money to provide Eaton with a good income. The Chief is anxious to see the matter disposed of but after having had two long interviews lately with Fred cannot see very much chance at present."

In February 1928, Van Norman traveled north to Long Valley to make a final attempt at procuring a sale on behalf of the city. George Khurts, acting on behalf of Eaton, had confidentially wired Mathews at the Cosmos Club that if Mathews would tell him the price the Board of Water and Power Commissioners was willing to pay for Long Valley, he, Khurts, would talk the ailing Eaton into it if the price were "not too low."

Van Norman's attempt was in vain. After months of effort, no agreement could be reached. Eaton stubbornly would not settle. Just as stubbornly, he managed to keep the embarrassment of his financial woes hidden from his friends and one-time admirers in Los Angeles. And for the time being, the intractable Fred Eaton disappeared as a topic of conversation in the offices of the Department of Water and Power.

13

RIVERS OF HADES

Sudden destruction
cometh upon them.
1 THESS. 5:3

ON MARCH 12, 1928, at four minutes before midnight, the world's safest dam, the St. Francis, stood in all its technically perfect engineered glory. Mulholland personally christened its completion in the summer of 1926 without fanfare due to the prevailing troubles in the Owens Valley. It was hailed as one of the most impressive designs of its kind anywhere. Mulholland had been lauded for his construction genius in its design.

At three minutes before midnight, the St. Francis Dam suddenly collapsed, spewing twelve billion gallons of water into the narrow San Francisquito Canyon at its base. At Powerhouse Number Two, one and a half miles down the canyon, engineer Ray Rising awoke to a deafening roar. He threw open the front door of his living quarters and stared in awe at a seventy-eight-foot wall of water crashing towards him. Rising yelled for his wife and three small children to run for their lives. Before they could escape, the water engulfed the concrete powerhouse and Rising's adjoining cottage, collapsing the wood-frame house like a box of matchsticks and dragging it along in its chaotic path down the valley. The force of the deluge smashed the power plant into ten-thousand-ton pieces of concrete and carried them along on its wild ride.

As the monstrous wave hit, Rising struggled to grip his wife's arm, only to feel her slip away. Then he felt himself enveloped in blackness and choking with mud, his lungs filling with water. Tossing him on his back, and then head over heels, stripping his body of all clothing, the raging water propelled him down the valley. Then, miraculously, caught in the swirl of a freakish eddy, Rising managed to grab hold of a section of rooftop, pull himself up on it and ride it spread-eagled until it smashed up against a hillside and hurled him to safety. Through it all, he shouted in vain for his wife and children. Vomiting water and dazed, all he could do was helplessly watch as the ten-story wall of death rolled on.

The raging head of black water, moving at eighteen miles an hour, now swept through Castaic Junction and boiled into the dry Santa Clara River bed, heading west. The water whipped through the small towns of Piru, Fillmore, Bardsdale, and within three hours flooded Santa Paula, forty-two miles south and seventeen hundred feet lower than the dam. The murderous wall of water now dwindled to twenty-five feet but it still splintered three hundred homes in Santa Paula's southern section. The death and destruction continued, and swirled through the hamlet of Saticoy then across the Pacific Coast Highway ten miles west into the Pacific Ocean between Oxnard and Ventura.

The torrent swept clean sixty-five miles of rich, fertile valley, engulfing automobiles, bridges, locomotives, everything in its path. One-hundred-ton blocks of concrete rode the water like rubber ducks. Ranch houses were crushed like eggshells, their cement foundations pulverized. Steel bridges were smashed like tin cans, and acres upon acres of citrus and nut trees were uprooted, their groves wiped clean of all vegetation. When it was over, parts of Ventura County lay under a seventy-foot-thick blanket of slimy debris.

In Los Angeles, Van Norman was notified of the dam's failure at 1:09 A.M. He immediately dressed and drove four miles in a light drizzle to Mulholland's home. Van Norman walked up the wet slate porch steps to the front door, and rang the door bell. The household lights switched on, and Rose Mulholland pushed her head through an upstairs window as a neighborhood dog barked loudly at the intruder.

"Van?" Rose asked, startled.

"Rose," he said quietly. "I must see the Chief." Within minutes, William Mulholland appeared in the doorway, dressed in his pajamas, slippers and silk robe.

White-faced, hat in hand, Van Norman looked at him for a long, painful moment.

"What is it, son?" Mulholland asked.

"The St. Francis is gone," Van Norman said.

Following Mulholland to his upstairs bedroom, Van Norman enumerated some of the known casualties. Rose quickly retrieved her father's heavy work boots and field coat from the upstairs closet and brought them to him. Instead, Mulholland asked for the formal clothes he would wear to a funeral. Mulholland's chauffeur was summoned and he and Van Norman were driven to the St. Francis at 2:30 A.M.

The roads in the path of the floodwater had been washed out, and Mulholland's driver was forced to steer the big black Cadillac up an unpaved fire trail to get to a viewing site east of the collapsed dam. Water to Mulholland was God's gift to humanity, and he had lived his life in service of bringing it to the people of Los Angeles. Now he was about to witness its awful power of destruction. Standing silently on a rise, Mulholland and Van Norman gazed out at the scene below them. They could see in its entirety the huge span of the Santa Paula Railway bridge wedged in a muddy stream, broken and twisted, a tangled mass of spaghetti-like steel. As far as the eye could see, the flood waters had obliterated every shred of vegetation in the San Francisquito Canyon, leaving nothing but exposed granite rock and a moon-illuminated surface of mud holes. Dozens of shadowy clusters of ranchers were moving in the pre-dawn through flooded fields, attempting to remove human bodies and livestock buried in the silt and mud.

Mulholland was overcome by grief, and Van Norman, fearing he might collapse, took his arm to steady him. The men made their way down a steep mountain slope through scrub brush where they could get a good look at the remains of the massive dam, and where they found department employees searching for missing workers. Every stick, stone and nail of the powerhouse, including its adjoining houses, had been demolished. All 74 residents of the dam, and 140 workers in a department camp downstream, had been lost.

By dawn, Mulholland and Van Norman stood among fifteen hundred area residents who had survived the flood, some still bleeding and half-dressed, at a gathering on surrounding hilltops to watch in horror the last waters of the great deluge receding into a brownish gray swath of mud three miles wide. Amid pitiful, heart-wrenching cries of despair, slowly revealed in a hellish scene were hundreds of dead loved ones—fathers, mothers, sons and daughters, brothers and sisters, and family pets, who were unable to escape the destruction, some of whom had been impaled by tree branches, others pinned between rocks and chunks of concrete.

Their faces gray with fatigue, survivors hunted for the missing while others collected their dead. Men sloshed precariously through quicksand-like muck, carrying makeshift stretchers and portable camp kitchens to aid and feed survivors stranded and shivering on small shoals in the middle of still-turbulent streams.

Few had any chance at all. Trapped below the steep canyon walls, the victims could only run a short distance before they were overtaken. Those who had paused to grab shoes or coats had been immediately swept away. Like Ray Rising, only miracles saved those who survived.

In many places the mud was fifteen feet or more thick, hindering the search, and a foul stench permeated the entire valley. Armies of volunteers arriving on the scene methodically fanned out over a distance of twenty miles, searching every crevice and beneath every boulder, poking long sticks into the mud, prodding for bodies. Rescuers then marked the locations of corpses with iron markers pounded into the mud. The work was gruesome and disheartening. Some bodies were so battered by the awesome force of the water that identification was impossible.

Teams of pack horses and mules from nearby ranches were used to drag the silt-covered bodies to dry land. Ambulances arrived in Newhall, and a makeshift morgue was established in the town's dance hall where bright decorations still adorned the interior from the previous festive Saturday night. By noon, undertakers could not cope with the mounting number of corpses, and fifty mud-caked bodies lay in rows on pine boards, piled like cord wood, and cleansed with garden hoses for purposes of identification.

Otto Steen, leader of a ten-member rescue team from the Department of Water and Power, himself located twenty-two more bodies near the shattered dam. Later, moving downstream, he heard a cry for help. It was Ray Rising, huddled naked in the scrub brush, cradling a matted sheep dog. They carried the exhausted Rising three miles to a Red Cross camp, the mud-covered sheep dog treading miserably behind, wagging his tail.

Families searched in vain while numbers of their dead grew—one hundred, two hundred, three hundred. They roamed the makeshift morgues and school gymnasiums throughout the area of devastation, searching for their loved ones. Forty-two children—half the enrollment of the small Saugus Elementary School—were dead. Some feared the death toll could reach as many as six hundred—many of the bodies had been washed into the ocean, and many missing Mexican migrant farm workers could never be fully counted.

Six hundred vehicles from the Los Angeles sheriff's and police departments and the California State Patrol descended into the flood zone. Their bright uniforms quickly turned a drab, muddy olive as they searched through the muck. Red Cross ladies set up camp kitchens and sleeping cots, brewed giant pots of steaming coffee and loaded hundreds of sandwiches into steel bins.

Fifty tractors followed by squads of recruited laborers moved down the Santa Clara Valley toward the Pacific Ocean in a frantic search for more survivors. At night, a reddish glare of gloom hung over the Santa Clara Valley as workmen set fire to huge piles of refuse. Like a somber funeral dirge, the hum of the tractors muffled the blows of axes and picks as workers continued the grisly search. An appeal for fifty more tractors was broadcast by Los Angeles radio stations, and volunteers were told to mobilize at what used to be a half-million-dollar showplace and tourist attraction, the ranch near Castaic of movie actor Harry Carey, now wiped clean.

Witnesses recounted to reporters freakish scenes of survival—a nude man seen floating on an empty wardrobe trunk, a woman riding the top of a city water tank in an evening dress—a mother and her three infants gratefully clinging to an old featherbed mattress miraculously lodged in a tree.

Tent cities were quickly erected in the flood zone. In Santa Paula, where the flood waters had reached as high as the second floor of the high school, fourteen houses were seen randomly floating like sticks of straw over the inundated Isabel school grounds. Twenty blocks of Santa Paula houses were destroyed. The hundreds of somber, mud-encrusted searchers and workers streamed through the makeshift camps seeking shelter, clothing, and news about the missing.

On a hillside near the Carey Ranch, a pathetic figure of a woman sat huddled only in a tattered, red, water-stained cotton sweater, wringing her hands, sobbing over her missing little daughter, her home now nothing but a mud hole, and the cottonwood trees stripped of bark by the scouring waters. Deputy sheriffs approached her from an islet in the receding waters, carrying the body of her three-year-old child. Over her nightgown, the child had a Sunday-best coat and on her feet her mother's shoes with the strings untied, mute testimony to the frantic effort the mother had made to save her. Behind the sheriffs walked two brawny, hip-booted ranchers in tears, carrying the remains of two other children.

At the Santa Paula Hospital, overflowing with victims, another mother, Mrs. Ann Holsclaw, wept as she explained how the flood current washed her home away and threw her to safety while simultaneously ripping her baby from her arms. "Johnny was sleeping with me," she sobbed. "I clutched him as tight as I could when the water swept us out in the dark. I grabbed hold of something. With my other arm I held the baby out of the water best I could, then I couldn't hold on any longer. Why was I the one who had to live?" she asked plaintively.

RECOVERING FROM THE initial sight of the immense human tragedy the flood had wrought, Mulholland and Van Norman joined in with the rescuers to aid them. Grief-stricken and fatigued, Mulholland returned to the offices of the Department of Water and Power eighteen hours later. His immediate concern now was to secure the city's water supply, and he ordered breaks in the aqueduct system caused by the dam failure to be repaired. Telephone linemen worked feverishly to restore communications to the devastated area while Department of Water and Power crews began to secure the broken aqueduct pipes, assisting where they could in the rescue of survivors.

At Mulholland's order, conscripted air mail pilots began an immediate air survey of the entire area from the Santa Clara Valley to the Pacific Ocean. At dusk Mulholland received the overwhelming details. "It's a scene of horror, Colonel," reported the weary voice of one of the pilots over the telephone.

"It is just one great scene of devastation . . . some places a half-mile wide, and at other places a mile and a half. It stretches clear to the sea. Thousands of people and automobiles are slushing through the mud and debris looking for the dead. Bodies have been washed into the isolated canyons. I saw one alive stuck in the mud to his neck."

The pilot continued with a grim prediction that he personally felt further rescue work would prove fruitless until sunrise as a dark, moonless night was anticipated. To attempt to rescue survivors trapped in the treacherous mud at night would only cause the death of the rescuers, and the bodies of the already dead, human and animal, would eventually be recovered in the huge mounds of tangled debris that had swept down and was now piling up at the ocean.

Desperately Mulholland desired to continue the search despite the advice that it would be fruitless and dangerous, and was about to proceed when Van Norman convinced him of the hopelessness of the attempt. For what seemed an eternity to Van Norman, Mulholland, defeated, sat down heavily in his desk chair and covered his face with his hands. Van Norman knew that for all his past glory and power and fame, his Chief, the conqueror of the Los Angeles Aqueduct and builder of mighty dams, the friend of governors and presidents, could do nothing but wait until dawn. And by dawn's first light, the state militia, the state police, and later the United States government, backed by three thousand volunteers, were beginning to arrive to search for the still-living under Mulholland's adamant direction.

While the search for human victims continued, the thousands of animals that perished in the flood caused serious health concerns for officials. Fearing an outbreak of typhoid, officers ordered human survivors inoculated with typhoid serum, and insisted on the cremation of all animal remains. Scores of marooned domestic fowl had to be removed from debris and trees to areas where they could forage for themselves. It was also necessary to slaughter thousands of hogs and other livestock because of the impossibility of transporting them through the treacherous mud. Those that were saved were enclosed by barbed wire fences to contain them. The barbed wire proved to be a source of injury, maiming many of the panicked animals as they attempted to flee. There was no choice but to kill the suffering animals.

News of the tragedy had now reached the nation. People who had never heard of the Santa Clara Valley now found their attention riveted on radio and newspaper accounts of the dam's collapse and the ongoing rescue efforts. The *Los Angeles Evening Express* was the first newspaper to publish the gory photographs of the disaster. Photo plates had been rushed by automobile after newspaper photographers invaded the disaster area. "Corpses Flung in Muddy Chaos by Tide of Doom" read the morning headline over the heartbreaking photographs. The next morning editors typeset a new slant for an eager and morbid audience: "Desolation Stalks Where Fertile Fields Once Held Happy Homes Now Hurled Into Oblivion" printed above photographs of mud-blanketed farms.

Seeking to view the carnage firsthand, thousands of curiosity-seekers descended in droves. By the weekend twenty-five thousand cars descended into what was left of the town of Santa Paula, and all major thoroughfares in the San Fernando Valley were shut down in the biggest traffic jam in its history. Newspaper and radio pleas urged motorists to stay away, which perversely excited their curiosity even more. Angry ranchers forced the curious and unruly mobs back with loaded shotguns. Police were issued orders to shoot looters on sight.

In the great flood's aftermath, debris, bodies, and pieces of dam were found washed up on the beaches of San Diego, two hundred miles south. The headless body of another victim was found in the river bottom near Santa Paula. Because of the battered condition of the torso, the Ventura coroner was unable to determine whether the victim had been a man or a woman.

The flood had cut a two-mile-wide path, stretching seventy miles from Santa Clara to the Pacific Ocean. At least 450 people were known dead, 1200 homes were destroyed and 8000 acres of farmland stripped clean. Damage estimates were reported at $15 million.

FIVE DAYS AFTER THE FLOOD, Rose Mulholland opened a letter addressed to her father written by Anna Holsclaw of Santa Paula.

Dear Sir,

Will you please read this letter thoroughly as it comes straight from the heart of a broken hearted and distracted mother who,

> through the St. Francis dam disaster has lost everything on earth including two darling children, one little girl twelve-years-old whose body we have recovered, but our darling baby's body six-months-old we have never found. . . .
>
> I will never know a minute's peace on earth till I have his little body and lay it tenderly and peacefully away beside his little sister, although we know their little souls are resting with Jesus in heaven. I am asking, begging and praying that you will help us in this search for our darling. . . .
>
> I will not give up hope, our darling baby lies buried in sand or debris and I must find him. Will you help us Mr. Mulholland? We have a clue and if you promise us either with funds or men and tractors to comb the river bed where we think our baby lies buried.
>
> You no doubt are a kind and loving father, do you remember when your children were babies? Can't you still feel their little clinging arms around your neck and their loving kisses?
>
> In God's name remember them and then think of me, empty arms, my broken heart and the long and lonely days and nights ahead of me, I'm not trying to blackmail you Mr. Mulholland, I have simply opened my heart to you asking for help, will you help me?
>
> The Bible says, "Ask and ye shall be forgiven, Seek and ye shall find."

After reading it, Rose sat down and wept. Her tears were a prologue for the many she would shed for her father in the coming days and months.

14

BREATH OF VENGEANCE

These be the days
of vengeance.
LUKE 21:22

IN THE IMMEDIATE DAYS after the failure of the St. Francis, there was scarcely a man or woman in Los Angeles who did not have his her or own theory on why it had happened. A minor earthquake in Santa Barbara three days before created speculation that the earth's movement had weakened the dam's foundations, which were rumored to be built on a fault line. Dynamite blasts by Department of Water and Power road workers near the dam the day before the disaster kicked off further speculation. Ten days before the break, residents living below the dam reported seeing what appeared to be leakage at the base. A theory offered by Mulholland supporters strongly suggested that the dam was sabotaged by vengeful Owens Valley insurgents.

But to many grim-faced Santa Clara Valley mothers and fathers who in the aftermath of the flood formed a seemingly unending parade through the makeshift funeral parlors and morgues, only one institution and one man was culpable—the Department of Water and Power and William Mulholland. These ranchers and farmers who had enjoyed one of the most bountiful areas in America were devastated, and feared that many of their kin and neighbors were forever lost in the wake of the watery holocaust. For the next fifty years there were periodic discoveries of grim remains that brought the death count to nearly five hundred.

The altruism that encouraged Santa Clara Valley survivors to help one another also fueled the growing anger against Mulholland and the city of Los Angeles, the "criminals" the residents believed responsible for their plight. The basis for their seething emotionalism had originated years earlier in sympathetic response to the Owens Valley water wars. To the flood survivors, the reasons behind the dam break were immaterial. That it happened was merely a predestined outcome of the devious and ruthless duo of Mulholland and the city of Los Angeles in their constant, greedy quest for water.

Twenty-four hours after the break, one incensed woman, who had lost her entire family, painted a sign which she hammered into the soggy ground before her destroyed home. On it in dripping red paint were the words, KILL MULHOLLAND!

Reporting speculation on the causes of the catastrophe in fervent editorials, disaster-mongering newspapers pandered to inquisitive audiences around the world. Attention was fixed on the tragedy not only because of the magnitude of the destruction, but also because there had been relatively few cases of big dam failures anywhere, and the collapse of the St. Francis, less than two years old, astonished engineers and laymen alike. Growing accusations that Mulholland was responsible became the topic of the day. The news columns of Board of Control member Harry Chandler's *Los Angeles Times* appeared to be sympathetic to the beleaguered Mulholland but were thinly veiled accusations that the city and the board's rapidly tarnishing hero were at fault.

"Mulholland's Heart Torn By First Disaster," read the *Times*'s subtle headline as early as the day after the flood, and reported:

> Chief Engineer William Mulholland was a pitiable figure as he appeared before the Water and Power Commission yesterday afternoon to make his report on the visit to the scene of the disaster. His figure was bowed, his face lined with worry and suffering. As he told the commissioners of his trip his voice was broken. Every water commissioner had the deepest sympathy for the man who has spent his life in the service of the people of Los Angeles, administered the Water Department from village days to the present and made the Los Angeles of today possible by building the Aqueduct and furnishing a supply of water for a city of 2,000,000 persons. In all his career of handling great projects he is facing the first disaster to any of his achievements. For his Irish heart is kind, tender, sympathetic, and the tragedy for the people in the canyon and the Santa Clara Valley is the tragedy of William Mulholland.

However, the ubiquitous Board of Control, knowing only a clean sweep of all those responsible for the building of the doomed St. Francis would pacify an outraged public, dropped the political hot potato into eager but far less sympathetic hands.

"Resign—Now!" demanded the muckraking *Los Angeles Record*'s blatant headline, sparing no words. "The St. Francis dam failed—and 400 people died—because of the engineer who built it . . . " led off the fiery front-page reproach that further demanded the resignations of R. F. Del Valle, president of the Board of Water and Power Commissioners and commissioners J. R. Richards, William P. Whitsett, John R. Haynes, and Will E. Keller. Calling for Mayor Cryer to kick them out by legal means and by force of public opinion if they did not quit on their own, the *Record* charged the commissioners had abetted in the disaster through their "stupidity, incompetence and arrogance." "They are the men who allowed William Mulholland, ignorant of modern, big dam engineering, to build this flood peril, without check or guidance from expert, disinterested engineers and geologists. . . . Mulholland must quit, but if he quits alone it will be a disgrace to the city of Los Angeles. It will be terribly unfair if he is made an official goat for the deadly blunders of his immediate bosses. . . ."

Like good salesmen, the Board of Control knew that its dazzling new product—the city of Los Angeles—had to be continually publicized in order to be sold. Clowns on stilts and nubile beauties in skimpy bathing suits lured in the crowds and sold neckties and automobiles. But to sell the fabled City of Angels, stupendous events were needed to continue to bring in the millions of eager-to-be homeowners seeking utopia in the rapidly rising housing tracts of the San Fernando Valley, their pockets burning with cash hard earned in the wheat fields, coal mines, and steel mills of America. The board was quick to realize that unlike the magnificent opening of the Grand Cascades and the flower-strewn dedication of Mulholland Drive, the collapse of the St. Francis was detrimental to their cause.

In a benefit for the victims of the disaster, a glittering array of movie stars appeared in a midnight gala at the Metropolitan Theater. One city block between Broadway and Hill at Sixth Street was illuminated by sixteen searchlights, donated by Adolph Zukor, president of Paramount Studios. Staged by Syd Grauman of the famous Grauman chain of theaters, the event featured six masters of ceremonies and a star-studded lineup of talent including Jack Benny, Gloria Swanson, Charlie Chaplin, W. C. Fields, Buddy Rogers, Fay Wray, Sammy Cohen, Glen Tyrone, Stan Laurel, Oliver Hardy, Irving Berlin, Jack Dempsey, Tom Mix, and the Ziegfeld Ingenues.

However, one important celebrity was missing from the dazzling affair. For the first time since the opening-day ceremonies of the aqueduct, William Mulholland's illustrious presence had not been requested to glorify a civic event by the city's fathers. Secluded in self-imposed isolation, Mulholland remained entrenched inside his St. Andrew's Place home. Listening to the NBC broadcast of the benefit, Rose Mulholland quickly switched off the radio when she heard her father's footsteps as he entered the living room. His face was lined with grief, as it had been in the previous nights since Van Norman first told him the news that the dam had burst. Since then, racked by a black, all-enveloping sorrow, he had not slept more than a few hours and Rose knew that tonight, despite the goings-on downtown, would be no different. If anything, news of the benefit only added to his despair.

Rose went to the kitchen and returned with his favorite late-night snack of cold roast beef and potatoes, and poured him a tumbler of whiskey. Ignoring the food, Mulholland took the whiskey and retreated into his wood-paneled study and closed the polished doors. He remained there alone with his despair until dawn. That night, Rose knew still another disaster lay ahead for her seventy-three-year-old father to bear. He had received a subpoena from the Los Angeles District Attorney's office; after more than fifty years of public service and national fame, the great visionary and engineer was now under investigation for manslaughter.

WITH UNPRECEDENTED SPEED, by the weekend of the break eight separate investigations were under way by city, state, and federal authorities to probe the causes of the disaster, prompting Los Angeles District Attorney Asa Keyes to announce his intention to file criminal charges against the person or persons responsible for the catastrophe. The legal battles surrounding the collapse of the great dam were set in motion, and a coroner's jury was scheduled to begin deliberations the following week.

For the time being, the facts surrounding the failure of the St. Francis Dam were now buried beneath tons of broken concrete and mud. As young Assistant District Attorney A. J. Dennison resignedly pointed out, the dam break and ensuing flood had committed the perfect murder—no living eyewitness and complete obliteration of the physical evidence. But his boss, Asa Keyes, thought differently. Hoping to add a final building block to his lofty political ambitions, Keyes moved quickly to pounce on the public's outrage over the St. Francis disaster, just as he did in the 1924 outrage over the aqueduct bombings. He and Mulholland were united against the Owens Valley insurgents, but now Mulholland was the accused and Keyes his righteous accuser. In the immensely adored William Mulholland, Keyes had a much larger trophy to add to his political mantel than the heads of a group of rag-tag dissenting farmers and ranchers.

The coroner's jury was instructed to merely ascribe the cause of death and attribute blame without legal effect, but Keyes knew these inquests were notorious for swaying public opinion, and the circumstances surrounding the collapse of the St. Francis with its bitter emotionalism and political fallout were ideal for exploitation.

The public was demanding a head to fall and Keyes would give them Mulholland's on a platter. Keyes's plan was to exploit the tragedy by attacking the aging engineer with the zeal of an overpaid press agent, all the while displaying his fearless district attorney's courtroom demeanor and his well-known penchant for expensive clothing.

Days before the inquest began Keyes staged highly visible press conferences blaming engineer Mulholland's incompetency and criminal neglect for the disaster. Creating further public outcry and panic, he insinuated that Mulholland's nineteen other dams and reservoirs were possibly on the verge of bursting.

In 1928, by law in Los Angeles, a coroner's inquest was required in all cases of death except those attributed to natural causes. "If it is possible for the jurors to place the blame for the collapse after they have heard all the testimony," Asa Keyes firmly told reporters, "I will demand that they do so, justice must be served!" On March 20, exactly eight expedient days from the collapse, a mob of eager spectators and witnesses surrounded the city of Los Angeles's ornate Hall of Justice and the nine formally dressed members of the jury entered the barrel-vaulted entrance foyer accompanied by deputy sheriffs. In the small Coroner's Inquest Room, the hundreds who poured in to watch the proceedings could not be accommodated, and once the last seats were filled, the deputies locked the doors.

Inside sat Harvey Van Norman, W. B. Mathews and several members of the Board of Control, and the Board of Water and Power Commissioners. According to the custom of the time, Bessie Van Norman, dressed in black, sat with the few women present in the back row. Only a handful of the dozens of journalists that had descended to cover the story could be accommodated, and the others stood noisily in the hallways with the crowd.

The solemn-faced jurors had been selected for their training and background in matters of engineering, science, and business, and were considered impartial. Keyes's booming voice opened the proceedings by calling the first witness, autopsy surgeon Jonathan Webb. Improving on the age-old adage that a picture was worth a thousand words, Keyes, to start the proceedings, had bailiffs push a gurney with a sheet-shrouded body into position where it could clearly be seen by jury members. The noisy chattering in the room suddenly fell silent.

"Did you, Dr. Webb," Keyes asked, pointing dramatically to the gurney, "make an autopsy on the body of Julia Rising?"

"I did."

"Will you state your findings, please."

"The autopsy was made on the fifteenth of March, 1928. The body was a female of the white race, aged twenty-nine years and five months, height five-foot five-inches, estimated weight 175 pounds, dark brown hair and light complexion. Examination showed numerous superficial scratches, punctures and bruises scattered over the body, the face and the limbs. There was a deep gash three inches long from the center front of the left leg and a laceration on the right forehead. The lungs were red, inflamed and contained water—the trachea and stomach were filled with a considerable amount of silt," declared Dr. Webb in grisly, impassive detail.

"And, Doctor, did you reach a conclusion as to the cause of death?" boomed Keyes again, his eyes now firmly fixed on the figure of William Mulholland seated among the witnesses.

"From these findings," Webb said, "the deduction was made that death was due to drowning."

Asa Keyes now called for a witness to identify the body. All eyes turned to the widower, Ray Rising. Dr. Webb pulled back the sheet from the face of the corpse and Rising beheld his wife for the first time since she had slipped away from his grasp in the deadly current. After identifying her, Rising testified that he was a city worker living with his wife and three children, Adeline, twenty-two months old, Dolores, three years old, and Eleanor, ten years old, near Powerhouse Number Two, one and a half miles below the dam. He had worked for the Department since 1923, planning to be a bureau employee until retirement. Rising sobbed as he recited his family's names, and bit his lip to fight back the tears.

Asa Keyes declared that the body of Julia Rising represented sixty-nine others from Los Angeles county who had perished in the flood. He somberly intoned into the record the long list of names of the victims and physical descriptions of two unidentified dead. He then asked for the next witness, calling out the name with a tone of controlled indignation. A murmur filled the room, as William Mulholland stood to be sworn in. Old friends and associates seated in the room whispered to one another that the Chief looked like he had aged ten years since the disaster.

Mulholland had already faced the grim reality of the morning with bitter news of the discovery of more bodies, those of a baby boy and a middle-aged woman still unidentified. Their discovery now brought the death total to date to 277, with an estimated 500 still missing. Observers noted that during Rising's testimony, Mulholland stared only at his thick, workman's hands folded in his lap.

Dressed in black mourning attire, double-breasted wool coat, high collar, and silk cravat, Mulholland unsteadily took the witness stand and, raising a trembling hand, swore to uphold the truth.

With the audience straining to hear, patiently, quietly, and directly he began to answer the many questions put to him by the sundry members of the inquiry. But thoughts of the magnitude of the great tragedy had overwhelmed him and occasionally his answers strayed into revealing self-pity.

"On an occasion like this, I envy the dead," he muttered almost inaudibly to no one in particular at one point during the proceedings as if to sum up the depths of his despair.

Mulholland's revelation of his pain had no place in Keyes's sense of compassion and he rolled his eyes mockingly for the benefit of the jurors at Mulholland's words. Seeing a clear-cut vision of a grandiose life in the governor's mansion on the Sacramento River, Keyes began his interrogation, or "inquisition" as Mulholland supporters reported later. With theatrics that even the most over-melodramatic silent movie star would envy, Keyes played to the jury and audience, punctuating each question and answer with facial gestures ranging from incredulous surprise to moral outrage.

Although Mulholland had enjoyed the immense fame that was heaped upon him since the completion of the aqueduct, he was by nature reticent to "flair in the limelight," and Asa Keyes's courtroom histrionics deeply galled him. All of his life, Mulholland had had a workman's distrust of "properly educated types." He despised men like Keyes who used their position as an elected city official to further their own gains, and like many others considered him a "scavenger." To Mulholland's engineering mind, Keyes didn't build or create anything; he only shuffled papers back and forth while languishing on the city payroll, using the misfortune of others as a rostrum for his own political agenda, all the while espousing advocacy for law and order.

Armed with a briefcase full of corroborating affidavits from key witnesses who were now ready to testify on behalf of the state that the dam as late as one day prior to the break was leaking, and leaking badly, Keyes was sure their testimony would prove the cause of the break. Department of Water and Power workers also knew of the leakage and this included engineers of the dam. More important, Keyes intended to prove that William Mulholland himself also knew of the leakage. In fact, Keyes would reveal that Mulholland was so concerned about the leakage, that after the dam keeper reported a new subsequent leak, Mulholland left immediately from Los Angeles accompanied by his aide Van Norman to inspect it himself on the eve of the break.

Mulholland denied none of this. He stated that he had been at the dam site just one day before the break on one of his routine checks, but had noticed nothing unusual. He testified that he made a habit of visiting each of his nineteen dams once every ten to fourteen days. However, the following day, on the morning of the break, dam keeper Tony Harnischfeger had telephoned him at the Department to inform him of what appeared to be muddy water collecting near a new leak at the end of the west wing of the structure. (Twelve hours later, Harnischfeger and his six-year-old son would be the first to die when the dam burst. Their bodies were never recovered.) By 10:45 A.M. Mulholland and Van Norman were at the site to inspect the seepage. Mulholland testified that to his relief they found the water clear. The leak was spilling down the slope of a hill and running across an old construction road cut into the side of the hill. The soft dirt from the road made the leaking water appear to be muddy.

Muddy water—that is, water containing silt or debris—leaking from any dam was a potentially dangerous situation, usually denoting earth or structural movement or both. But leakage in itself was not dangerous if it was clear. It was, of course true, declared Mulholland, that prior leaks were evident at the St. Francis and equally true that no dam, large or small, was immune from them.

In January, the typical spring run-off reached the St. Francis Dam, and Mulholland prepared for it by allowing the reservoir to fill to capacity. The water reached an elevation of 1,934 feet, only three inches below the spillway; by March 7, no additional aqueduct water was diverted into the reservoir. Leaks that had been evident the year before had begun to discharge once again, and new leaks suddenly appeared on both abutments. During the first week of March, a substantial leak developed along the wing dike releasing roughly .60 cubic feet of water per second. Mulholland had ordered his men to install an eight-inch pipe underdrain in this location eastward along the base of the dike so that water would discharge along the dam's western abutment.

By March 12, the reservoir was filled to capacity, and water was lapping over the edge of the spillway by wind-driven waves. All of

Mulholland's downstream storage facilities were full and excess water from the aqueduct was released into the San Francisquito Canyon the morning of March 12 for the first time in nearly two years. Water surged from the aqueduct's control gates at Drinkwater Canyon several thousand feet below Powerhouse Number Two. Residents who lived down-canyon and saw the water in the typically bone-dry creek bed stated later that they wondered whether something might be wrong with the dam.

"Like all dams, there are little seeps here and there and I will say, as to that feature of it, of all the dams I have built and of all the dams I have ever seen, it was the driest for a massive dam of its size I ever saw in my life," said Mulholland resolutely. "The water was clear . . . as clear as glass," he added.

Unfazed by Mulholland's insistence on the leak's benign character, Keyes, aided by Assistant District Attorney A. J. Dennison, pressed on with his customary prosecutor's zeal, zeroing in on the issue of leakage, looking for the slightest sign of malfeasance and incompetency in every innocent word, hesitant pause, or random cough. For the most part, Mulholland's answers were forthright. But the demeanor and brilliant repartee that had brought audiences to their feet in applause on the lecture circuit were subdued by the constant memory of the tragedy, and often during the long nine-day ordeal of the inquest, he found it difficult to concentrate on the questions put to him. With an enigmatic smile on his lips, he would sit staring silently at his hands until prodded by the questioner to answer.

But Mulholland was clearly a man under siege, and as the inquiry progressed, he was bombarded by Keyes and jurors with minute questions of the rock and soil composition on which the dam was constructed, the grade mix of the concrete, the reasons why the San Francisquito site was selected, the time it took to construct the dam, the depth of the foundation into the subsoil and bedrock and methods of drilling and coring, the anchoring of the dam wings to the walls of the canyon, the various types of earthen and concrete dam construction, the pressure of water against the structure, the leak frequency, the process of draining off surplus water reserve, the method of siphoning water from the aqueduct into the dam, and Mulholland's responsibility and manner of supervising the construction during the actual building phase and after. Charts, diagrams, and pre-break photographs of dam leakage were shown, and exhaustively analyzed. Rumors and statements of culpability were voiced and answers demanded.

To Keyes's accusation that Mulholland and the Department somehow knew of the impending disaster, no doubt caused by the leakage, and did nothing to save lives, Mulholland responded vehemently, shaking his head in denial of any prior knowledge or apprehension that the dam was about to give way.

"No, no, no! I—we had no more idea of danger, no more reason to believe there might be a catastrophe than a babe in arms. Man alive, I would have sent a Paul Revere alarm ringing through every foot of that valley. I would have been the first, the very first, sir, to spread the alarm. I would have exerted every effort to get every man, woman, and child out of the path of those terrible waters," he protested indignantly, adding that even if there had been many days prior notice of danger, it would have been impossible to eliminate the hazard. Opening all the flood gates would have reduced the water level only an inch or two a day. It would have taken months to empty the dam in safety.

He also hotly branded as false the rumors that Santa Clara Valley residents and employees of the dam threatened to move away because of any present danger. "Absolutely no such rumors ever were spread through the valley," he said. "They would have told me if they held any fears."

Asa Keyes stepped back briefly from his pungent cross-examination and shrewdly projected graphic motion pictures of the devastation on the bare white wall of the courtroom. The crude black-and-white film taken by a free-lance cameraman immediately after the catastrophe stunned the jurors with their first view of the ruins of the Santa Clara Valley. Weary men in heavy overcoats were dragging bodies to dry land, while others carried survivors on their backs. Corpses were strewn along the edge of the receding flood waters, animals were buried alive, there were huge oaks cleft above their roots, and automobiles washed downstream and half-buried in sand and mud. The pictures had their desired effect and the jurors recoiled in horror at the destruction. But it was Mulholland on whom the pictures had their greatest effect. Watching the scratchy gray scenes of death flickering on the wall, the gruesome scenes that he had been struggling so hard the past days to obliterate from his mind, he turned his head away, his sobs muffled by the steady whir of the film projector droning in the cigar-smoke-filled room.

To Keyes's chagrin, the vulnerable Mulholland, despite Keyes's aggressive attempts to place blame on him for the disaster, had successfully warded off Keyes's inferences and accusations of guilt. Now like a hungry predator lurking in high grass for wounded prey, Keyes chose the precipitous moment and instructed Assistant District Attorney Dennison to resume the attack.

"Mr. Mulholland, did you know when you erected this dam that it was in a fault zone?" led off Dennison.

"Of course. I've already testified to that," replied Mulholland irritably.

"Please tell us again."

"Well, you can scarcely find a square mile in this part of the country that is not faulty. It is very rumpled and twisted everywhere. I have dug underground, I suppose, seventy-five . . . one hundred miles of tunnels, and everywhere, there are faults and slips and crumples without exception. Look out of the window here at that bank, and you will see a formation that once was laid down flat by nature," he said, gesturing to a small window and the weedy rise seen beyond. "It was formed at the bottom of deep water and is now tilted up at about thirty degrees. The same formations are found all over the country."

"Then, certainly, it is necessary to build these dam structures so they will stand considerable horizontal pressure?" said Dennison, looking over as Keyes nodded approvingly.

"No, that is allowed for in the weight of the water," replied Mulholland.

"I mean in the fault zone?" quickly clarified Dennison.

"The foundation was not buried that deep in the ground. I have told you that before," retorted Mulholland testily.

"But it did rest on earth, did it not?"

"Yes."

"And was anchored into the hills on each side?"

"Yes, of course. It was subject to some stress there but nobody had ever been able to compute those stresses."

"And if it was in a fault zone—"

"They are all in a fault zone. The City of San Francisco reservoir is in a fault zone. The Haiwee Reservoir that supplies this town lies in a fault," he replied, with voice raised in exasperation.

"And engineers have to build them so as to make them fault-proof, don't they?" continued Dennison, eliciting a furtive OK sign from his beaming boss.

"They try to," Mulholland replied.

"And they do it?"

Still staring at his hands, Mulholland replied plaintively, "I have built nineteen dams and they are all fault-proof. I have built more than any other engineer who has testified here. And I have been consulted for at least that many more."

"I see. And all of them fault-proof? That's interesting," retorted Dennison sarcastically, and gave way to Keyes, eagerly stepping forward to take over the questioning.

"You have been an engineer, have you not, for thirty years," began Keyes with wide-eyed innocence.

"Fifty," corrected Mulholland.

"Well, that's a long time, fifty years. A long time. And, as you have said, you have erected a great many dams in that time?"

"Yes sir."

"Let me ask you then, as an engineer for fifty years and a builder of many dams, and an expert in faults, this fact: if water was leaking between the concrete dam foundation and the rock supporting it or between the wings of the dam and the rock conglomerate of the hills they were anchored into, would there not be a natural erosion and an absolute certainty that the dam would go out as it did?" Keyes asked, smugly confident.

"No. It might not go out at all. Just an ordinary crack in the material will cause it to respond to erosion, but wouldn't necessarily affect it severely," replied Mulholland defiantly.

"But if water in those areas was pouring sufficiently or cutting sufficiently, as persons in this room are prepared to attest, it would effect erosion and cause the dam to break absolutely, am I right?"

"That's correct . . . if it was sufficient. But I have told you it was not. . . . The water was clear."

"I see. . . . Well, I wouldn't know anything about that, but you would, of course, being an engineer for fifty years, and all. You certainly couldn't make a mistake about that, could you?"

Affronted, Mulholland glared, responding with: "I am here to give you all I know, and I swear to God that my oath is binding on me as—"

Keyes waved a dismissive hand in the air, bluntly interrupting with: "We are all aware of the nature of oaths here, Mr. Mulholland. It's the nature of dams we want to know about."

Then, with a mocking look to the jurors, Keyes added: "Tell us, if the leakage didn't cause the break, or a fault didn't cause it, have you any explanation as to the real cause of the failure of this dam? Certainly it didn't happen all on its own, or did it?"

Suddenly, without waiting for reply, Keyes grimaced and looked away in anger, only the anger was directed at himself. Asked for the sake of sarcasm and to solicit laughter, the grandstanding Keyes regretted the question the instant it came out of his mouth. Unwittingly, stupidly, he had just asked the very same question that he cautioned Dennison only hours before not to put to Mulholland under any circumstances. Such a question, Keyes knew, would lead the prosecution into the perilous territory it had wished to avoid at all costs.

If the question caused Keyes to want to bite his tongue in regret, it prompted Van Norman, the Department of Water and Power Commissioners, and the whole assembled room to eagerly lean forward in their seats in anticipation. The room was as quiet as death as they waited for Mulholland's expected reply.

Once again Mulholland glanced down at his hands. Pausing for what Van Norman thought was an eternity, he finally looked up and without taking his eyes off the small window, replied guardedly, almost reluctantly, as if wrestling with a great weighty decision.

"I have no explanation that could be called an explanation, but I have a suspicion . . . I don't want to divulge it even . . . It is a very serious thing to make a charge—to me it is a sacred thing to make a charge, even of the remotest implication. . . ."

At the word "charge," Keyes immediately knew he had in fact entered that perilous territory, and the lofty confidence that he was so ably demonstrating suddenly was pervaded by a sense of panic which he quickly struggled to control.

Now, as the obviously hesitant Mulholland sat before him in the witness chair of the inquiry room, Keyes's previously successful struggle to suppress the mention of dynamite was in danger of blowing up in his face, and ironically, as he very well knew, he had lit the fuse himself. Wiping the tiny beads of sweat that had appeared on his brow with a silk monogrammed handkerchief, the ever resourceful Keyes moved to correct his error. "Is that all you have to offer, a bare suspicion, Mr. Mulholland? Well, well, if it's only a suspicion it can't amount to anything, can it? I must remind you, this is a very important matter to everybody here," he chided as if correcting a small child.

Mulholland sighed heavily. "Yes sir. And it is most important to me. Human beings are dead."

"Of course, of course. But let us continue," replied Keyes dismissively. "Let me now ask a question from an engineering standpoint: was it not an utter impossibility to build the dam with any factor of safety in the manner in which it was constructed?"

"An impossibility?"

"Yes. To build it with any factor of safety—was it immune from failure?"

"If I thought that was the case I would not have built it."

"A great many men make mistakes and everybody is liable to make a mistake. All of us here in this room certainly understand that, Colonel Mulholland," Keyes said, patronizing Mulholland with a sweeping gesture to the audience. "But isn't it a fact, established by the best minds of engineering, that it would be an utter impossibility to build a dam there with any factor of safety?"

"I would not have built it if I thought that," repeated Mulholland, his attention drifting again to the window.

"That's what you'd say then, but what would you say if you had to do it over again—would you build that dam in the same manner in which it was constructed?" asked Keyes impatiently.

"No, I must be frank and say that now I would not."

Keyes took an audible breath, and sensing victory, pressed on. "And why wouldn't you build it there again, Colonel—because of these leakage factors we've just spoken of, correct? Isn't that it?"

"It failed, that's why," replied Mulholland. "There was a hoodoo on it. . . ."

"Hoodoo? Hoodoo?" repeated Keyes. Turning to the jurors he shrugged, opening his palms in a gesture of incredulity, then faced Mulholland again. "Tell us more about this 'hoodoo,' Colonel."

"Well, it was in a vulnerable spot. The break has all the appearances of a hoodoo."

"You don't mean it's hoodooed in that the tragedy occurred on the thirteenth of March, do you?" asked Keyes now shooting an incredulous look to the jurors.

"Well, I hadn't thought of that," mused Mulholland, shrugging his shoulders.

"Now that's as good a theory as we'll ever hear here," smirked Keyes to a ripple of laughter.

There is no evidence that District Attorney Keyes was a religious man, but his associates knew he had reason enough to get down on his hands and knees and give thanks to the patron saint of public prosecutors. Just minutes earlier he had been in danger of losing the case of a lifetime. Then, as miracles often do, one came inexplicably to his rescue in the form of a hoodoo backed by a recess for lunch.

WITHIN HOURS of the dam's failure, behind the closed doors of the Department of Water and Power and at City Hall, there had been a greater and deeper panic than that inflicting Asa Keyes as he stood pondering his next theatrical maneuver. The crisis was no laughing matter; the Department and Mulholland believed the dam had been sabotaged. Early confidential memos detailing Department response in the event of terrorism were abruptly put into force. All roads in the vicinity of the aqueduct and city reservoirs were ordered closed, floodlights were utilized at siphons, guards were posted, and automobile patrols dispatched. Mulholland ordered five hundred high-powered rifles delivered to dam keepers and guards throughout the Los Angeles water system. At Mulholland Dam above the Hollywood Hills, armed guards paced the reservoir perimeter twenty-four hours a day. Mayor George Cryer and other city officials feared for the safety of the city. The governor's office was notified, and appeals for federal assistance should it be needed were sent.

The next day, newspapers printed jittery accounts of "startling new evidence" suggesting that the St. Francis had been dynamited, confirming the worst fears of water officials. The evidence consisted of a newly frayed rope found dangling from a bush on a cliff above the St. Francis, discovered by Department engineers and believed to be identical in thickness and texture to a rope found in an aqueduct bombing of 1927. Department officials speculated that the rope was used to lower boxes of dynamite into the dam's foundation walls. And in Hollywood, an anonymous man discovered a crude discarded map of the St. Francis Dam the day after it burst. The map contained handwriting allegedly identical to that of one of the insurgents involved in the dynamiting of the No Name Canyon Siphon. The rope and map were delivered to Asa Keyes's office to be introduced as evidence at the Coroner's Inquest.

Further, a wooden crate stuffed with soggy dynamite was found by a truck driver in the Santa Clara wash, two miles south of the failed dam where it had apparently been carried by flood waters, and police were tracing its origin.

"If that dam was dynamited deliberately, it was conceived by a soul ten thousand times blacker and infinitely more cruel than the one that carved up the body of little Marion Parker," a water official stated, referring to the gruesome murder of a small child that had stunned Los Angeles citizenry. "And if it is proven that the dam was dynamited, then there will be recorded in history the most dastardly crime of this century."

15

PERSECUTION

Persecuted,
but not forsaken.
2 Cor. 4:9

It is understandable why Mulholland and the Department of Water and Power suspected that sabotage caused the disaster. Coupled with the "new evidence," the bursting of the St. Francis had occurred on the heels of repeated assaults on the Los Angeles Aqueduct. The bomb blasts of the No Name Siphon and others in the years that followed had interrupted the flow of water to the city, and water officials felt the St. Francis break was an extreme extension of the same violent pattern.

Months of investigation by Los Angeles police officials eventually uncovered what was thought to be the central ring of the Owens Valley dynamiting conspiracy. Produced in court in the Owens Valley town of Independence were two pieces of evidence against the dynamiters: dynamite powder packing crates and a confession by

Bishop citizen Perry Sexton. The arrest of Sexton and the six other suspected conspirators known as the "Inyo Gang" culminated in a preliminary hearing that was in progress on the day of the St. Francis collapse. Despite dramatic testimony regarding details of the aqueduct bombing, the charges were dismissed several days later due to lack of evidence. When asked for a statement by reporters, Mulholland, disgusted at the dismissals, retorted that the only statement he felt like making could not be printed.

Eleven months before the St. Francis break, the Los Angeles sheriff's office had responded to anonymous phone calls reporting that a carload of men were on their way from Inyo County with the intention of dynamiting the St. Francis. Within minutes, Department of Water and Power personnel and sheriff's deputies sped through the San Fernando Valley to apprehend the saboteurs, but they never materialized. Weeks of intensive investigation followed but no arrests were made, and no signs of conspiracy were uncovered. Investigators filed additional reports stating that they overheard threats by Owens Valley citizens that if anyone were to be prosecuted for the dynamitings, the city of Los Angeles would be sorry.

Respected city councilman Pierson Hall took the accusations against the bombing suspects so seriously that he immediately embarked on his own crusade to determine if the facts could support a claim of dynamite as the cause of the St. Francis disaster, and contacted the Hercules Powder Company in San Francisco to obtain reports on the effect of dynamite on concrete. Hall met with a nationally prominent explosives expert from Texas, Zatu Cushing, who had over twenty-five years of experience in the field. He was so convinced that the dam was dynamited that he paid his own expenses to travel to Los Angeles to investigate. After examining blocks of cement remnants found behind the central piece of the dam, Cushing flatly stated that, in his opinion, the structure had to have been dynamited. The effect of dynamite on concrete was the "same as weather on a log which lies out in the open. It will sound and ring under a hammer at first, but with time it will crumble so that it can be scraped off with the claws of a hammer, or if left long enough, it can be clawed off with one's bare fingers," he wrote in a report. Hall contacted other experts who independently verified this conclusion.

Hall and Mulholland speculated that the task of placing dynamite in the foundations of the dam would have been relatively easy. One individual could have quickly done the job. The San Francisquito canyon had been a popular camping spot, and hundreds of automobiles regularly traveled up the road near the dam without suspicion. The shock of a dynamite explosion would have shaken loose the dam's wing foundations, and, once loosened, would have allowed the tons of rushing water to tear through the opening and cause the dam to crumble under the pressure. The explosion would have been heard only by a person or persons living immediately below the dam. However, no such witnesses survived the flood, and Hall believed that whoever planted the dynamite on the precarious slopes of the dam canyon might have also perished with it.

At noon on the day after the break, Dr. Elwood Mead, Director of the U.S. Reclamation Service, arrived in Los Angeles to begin the official federal inquiry on behalf of the Los Angeles City Council. He was immediately briefed on the situation by Hall, and driven to the site of the failed dam. Mead and Hall were amazed to see the central, seemingly most vulnerable section of the great arched dam still standing, having miraculously resisted the twelve billion gallons of raging water that had destroyed everything else in its path. The sight lent credibility to Mulholland's theory that the wings of the structure had been dynamited simultaneously, leaving the center section standing intact.

The strongest and perhaps strangest evidence supporting sabotage was submitted by an investigating Stanford University zoologist, Edwin C. Starke, who reported finding no live fish in the eddies below the fallen dam. Instead, he was shocked to find innumerable dead fish above the dam, which he theorized had been catapulted there by a colossal explosion. When the fish were dissected, they were found to have ruptured lungs—burst by concussion.

The fact that no live fish had been found below the dam in the pools of water formed by the flood incited more than passing curiosity. But naysayers argued that the fish died not from a dynamite explosion but from the heavy, silt-laden flood waters. The "Stanford Fish" theory was refuted by newspaper editorials in Owens and Santa Clara Valley newspapers, and the *Los Angeles Record* declared, "We are inclined to believe that there may be a dead fish involved in the matter—a fish so dead that it smells to high heaven. And we think that this fish may be a red herring the Water Board would like to drag across the trail that leads to those responsible for the Saint Francis Dam disaster."

Affected by this and other strong public opinions against the dynamite theory, and fearing reprisals from the Santa Clara Valley, frantic department officials refused to comment further on the evidence or lend credibility to investigations of sabotage.

While Mulholland busied himself with the defense of the city's water system against the alleged saboteurs of the St. Francis, Water Department bureaucrats in closed session acted with undue haste under internal pressure from city officials including members of the Board of Control to dismiss him from the department and relieve him of his official responsibilities. Rising to Mulholland's defense, Van Norman pleaded passionately against dismissal; ultimately outnumbered, he left the room in tears.

But Pierson Hall, presiding chairman of the Water and Power Committee of the city council, and a stalwart Mulholland supporter, spearheaded a counterattack to save Mulholland and implored the board to reconsider, since formal blame had yet to be fixed. "In my view nobody is in a position to know what caused the failure of the dam. City, state and engineering societies are forming committees to find out what caused the failure. Bill Mulholland has dedicated his life to the city of Los Angeles in securing and maintaining a sufficient water supply and he is known throughout the world. Every single part of this great utility is a creature of his brain. It seems to me it would be precipitate to take such drastic action at this time."

So passionate was his plea that the board reconsidered, and in a complete about-face unanimously voted against firing Mulholland. Members also turned down Mulholland's request for a leave of absence so as to avoid any further embarrassment for the department. Mulholland remained for a time, in an excruciating dual role as leader and accused.

The lack of support for Mulholland was due to serious anxieties on the part of beset city leaders. At all costs, city leaders wanted to make peace with the still-angry residents of the Owens Valley and with the now-devastated Santa Clara Valley. They desperately desired to calm public fears about the potential threat of future terrorism, as wealthy Los Angeles boosters and the Board of Control wanted to promote a continuing influx of people into Los Angeles. And perhaps the most important reason was to safeguard the impending Colorado Dam bill that was now on the verge of passing and would seemingly guarantee for all time the future water supply for Los Angeles. If Mulholland's enemies could destroy the much smaller St. Francis at such a terrible price in lives and money, what unimaginable havoc could they bring to the Goliath Boulder Dam? To discover that the St. Francis was dynamited would vindicate Mulholland, but it would also surely halt or delay legislation for Boulder Dam.

When questioned by the district attorney's office as to why the evidence pointing to dynamiting—the map and the tell-tale rope—were not disclosed sooner, city and Department of Water and Power officials responded with caution. "The city has no desire to escape any moral responsibility," one department spokesman stated, "and you can readily see that if the city admitted being in possession of strong evidence of dynamiting, the public, particularly those people in the devastated area, might receive the impression that we were trying to dodge just claims against us, or to whitewash probable fault in construction."

TO ASA KEYES, the words "probable fault" translated into the name William Mulholland. To protect the fair and all-important name of the city of Los Angeles, another fair but lesser name would have to be defiled. Devising a whitewash of their own and placing blame on Mulholland, Asa Keyes and city leaders may have acted to thwart a thorough investigation.

"As far as I know," Asa Keyes told reporters outside the Hall of Justice, "there is no evidence that the St. Francis was dynamited." Despite the availability of numerous witnesses who could testify as to their knowledge of facts regarding sabotage, Coroner Frank Nance reacted with the same closed mind. "I have no knowledge of any evidence found in support of the dynamiting theory. My investigators have not made any such report to me."

No less than five individuals were near the dam within one hour of its failure and were available to testify as to the apparent conditions at the dam site on Monday evening. All five were department employees stationed at Powerhouse Number One, five miles above the dam. Only three of these five witnesses were subpoenaed by Asa Keyes to testify at the coroner's inquest.

Elmer Steen and Katherine Span had left their friends' home at the doomed Powerhouse Number Two at approximately 11:35 P.M., driving up the canyon's southeast wall and passing the dam's eastern abutment at roughly 11:45. They testified they had not seen "anything unusual," but commented that the dam was "quite spooky" and "terribly quiet" in the moonlight as they drove along the unpaved San Francisquito Canyon Road.

Within minutes, Ace Hopwell, who also worked at Powerhouse Number One above the dam, drove up the canyon by motorcycle, sometime between 11:50 and 11:55 P.M. He recalled seeing headlights up-canyon, presumably those of Steen's vehicle, but also noticed nothing unusual. Slowly, Hopwell climbed the graded road up past the dam. Only the high, dull glow cast by the moon hidden by the hills, Hopwell recalled, and the car headlights up the canyon broke the monotonous blackness. One mile above the dam, however, Hopwell stopped suddenly when he heard a loud, ominous noise. Sensing an unusual sound or shaking, Hopwell pulled over, dismounted his bike, but kept the engine idling, and lit a cigarette while he listened to strange crashing sounds in the distance. The noise, he assumed, was a mile behind him. Hopwell testified that he thought the sound was a landslide, a frequent occurrence in the area; the rumbling sounded like rocks rolling down the mountain. Assuming the landslide was behind him, Hopwell remounted his motorcycle and drove up the canyon reaching Powerhouse Number One. It was there he learned of the disaster. Hopwell was the last person to see the dam before its collapse and live to tell about it.

Two other employees from Powerhouse Number One were near the dam the night of the failure, but these individuals were never called to testify, and it is unclear whether Keyes deliberately excluded their testimony or whether he was unaware of their personal knowledge. These witnesses, who later refused to be identified, told a journalist that they had driven the San Francisquito access road between Powerhouses Number One and Two near the time of the dam's collapse. They had observed that "the road had dropped at least twelve inches, just upstream of the dam's east abutment," and, with "bated breaths," the party rode over the "danger area and back onto firm roadbed." This eyewitness account, never heard by the coroner's jury, would become critical in a later investigation.

By Wednesday, March 21, the second day of the inquiry, sixty more witnesses had been served with subpoenas to testify before the coroner's committee. Despite Keyes's strenuous efforts, public speculation regarding sabotage soared. Mayor George Cryer said he was suspicious from the outset that the dam had been tampered with. "I personally inspected the dam sixty days ago and everything was found to be in ship-shape," Cryer told reporters. "There was no hint of trouble up there and a man's got to be plain pig-headed not to realize the thing broke at the zero hour when criminals were likely to work." The mayor became the first elected Los Angeles official to publicly endorse the dynamite theory, joining the ranks of Mulholland intimates Pierson Hall and Harvey Van Norman.

In retaliation, the anti-Mulholland forces acted, branding the dynamite theory as a "ton of hullabaloo." "If a dynamite blast had destroyed the dam, surely there would be more physical evidence than merely a piece of rope, a scrap of note paper and numerous dead fish," they asserted in their newspaper articles.

The collected evidence remained in the hands of Asa Keyes who now denied its very existence, telling associates that the coroner's jury would never see it. However, newspapers had already reported the discovery of the evidence, quoting one handwriting expert that the writing on the map matched that of one of the Owens Valley insurgents involved in the 1927 bombings. Nevertheless, Keyes stoutly denied to the press that he ever laid eyes on the rope or map. "It's just rumor, call it 'hooey,' as far as we're concerned."

The dynamite theory was further squelched by statements made by Ventura County Under Sheriff Eugene Biscaluiz, who hotly telephoned the *Van Nuys News* to deny any truth to the rumors that dynamite caused the St. Francis collapse. "I know nothing of these rumors," Biscaluiz declared. "I know what I'm talking about, there is nothing to report. We are not here running down dynamite clues, but doing the same thing that we have been doing since the days of the flood, recovering bodies and taking depositions to introduce at the inquest."

Vehemently anti-dynamite factions accused "special interest groups" of planting the "plot evidence" at the dam site in order to lead the coroner's jury astray in fixing blame for the disaster.

So thorough was Keyes in suppressing the dynamite theory that only one of the 120 witnesses called before the inquest eventually testified about the missing evidence. Los Angeles Deputy Sheriff Harry Wright told jurors that no importance could be attached to the map found in Hollywood. "It might have been dropped there out of the pocket of some man who had been drawing something for his own amusement," the deputy swore to Keyes's satisfaction.

The infamous map had remained in the hands of Hollywood police for some time after it was first submitted by the "concerned citizen" who had found it. Then it was handed over to officials at the Department of Water and Power where it was photographed and examined by Mulholland and Van Norman. The Department held it for twenty-four hours then passed it on to the County Sheriff's office. After that, its whereabouts became unknown.

Shocked that the incriminating evidence never found its way into the coroner's inquest, insiders at the Department of Water and Power moved quickly to protect their legal interests, thus setting off a media war of contrasting viewpoints. J. R. Richards of the Board of Water and Power Commissioners made a formal demand for a grand jury investigation. "We do not know if explosives were used," Richards said. "Nobody can say definitely at this time, but we believe the matter should be investigated thoroughly and this includes the map, the fifteen feet of rope and the new affidavits gathered by the department of water and power in the last few days."

Upon reading Richards's demands for the grand jury investigation in the *Los Angeles Times,* Keyes threw his newspaper down in disgust. After breakfasting at the Beverly Hills Hotel, a short distance from his posh new home on Rodeo Drive, Keyes telephoned his office, and grilled his assistant Dennison for almost twenty minutes about the nature of the affidavits.

Reporters had obtained copies of the affidavits which apparently strongly substantiated the presumption of terrorism, and printed whole sections of the text in evening editions. One affidavit filed by an employee of an Owens Valley utility company described an incident on March 4, as he was walking toward his car on the main street of Bishop.

> As I got to the rear of my car I heard someone call from one parked near my machine to a man sitting in another car. The person calling said, "Carl, come here."
>
> The man addressed got out of the car, approached the car next to my machine and put his head between the curtains. A man's voice within said, "Carl, do you think 150 sticks of powder is sufficient?" The man addressed as Carl said, "Yes, that is enough if you use it right."

Another man in the car said to Carl, "How are we going to get them all to go at once?" The first man in the car then said, "We have electric caps that will set all the charges off at the same time, regardless of where they are. Jerry has both kinds. Which side will be best?" The other man in the car said, "I think the west side," and Carl said, "No, both sides."

The affidavit went on to offer a physical description of Carl, and added that the voice was familiar—the same voice the employee had once heard complaining about having been forced to sell his ranch to the city of Los Angeles for an unsatisfactory price. The employee did not offer a name for the man whose voice he recognized.

A second affidavit from an Owens Valley resident described a hitchhiker who had made violent threats against the department. "At the time we were passing the aqueduct siphon in Mint Canyon my companion said things to me like: 'I would like to tear into this siphon. I would like to blow this and that out.'

"I said to him if he did that it would drown lots of innocent people, and his answer was, 'What do we care if we drown half the people of Los Angeles? We are going to show them we have not really commenced yet. Four plants of dynamite charges are ready where they will do the most good, and this time there will be bloodshed.'"

Suspicious of the authenticity of the affidavits and affronted by the one-upmanship of the Mulholland forces, Keyes, acting quickly but in clear violation of the public trust, decided not to present the affidavits, claiming the documents were hearsay and irrelevant. Because the proceedings were a coroner's inquest, not a full-blown trial, Mulholland was not represented by a lawyer of his own. Not only were the affidavits not shown to the jury, but some of the most compelling evidence which tended to corroborate a sabotage defense, namely the "Stanford Fish" theory and Pierson Hall's convincing evidence from the dynamite experts, was never presented for examination.

OF COURSE, to have pressed Mulholland further on the dynamite issue would have foolishly courted disaster for Keyes's case for leakage, but the reasons behind Mulholland's painful reluctance to assert terrorism as the cause were as complex as the man himself. Although he had privately espoused the dynamite theory among water associates as fact and had immediately acted to protect the city's water system from more violence, Mulholland had discovered in the witness chair that he could not publicly admit to it, even though to do so might have allowed the dynamite evidence into the inquiry and vindicated him of charges of incompetency.

It also would have left him defenseless in the face of the enduring accusations that the Owens Valley water wars were brought on by his unyielding dictatorial refusal to purchase Fred Eaton's Long Valley property for the city's critical water storage facility, and that his failure to come to a fair settlement with the displaced Owens ranchers created the hostile climate that prompted the dynamiting of the St. Francis by still-vengeful insurgents. A new war would then break out in the Owens Valley, only this time families of the Santa Clara Valley victims would be the ones seeking revenge. A protracted wave of bloodshed would shake both valleys and Mulholland would be held accountable.

Branded the villain in either case, Mulholland knew he could not win. Allowing the dynamite issue to disappear amid the mocking tones of Keyes's repetitions of the word hoodoo, Mulholland would continue to fight against the charges of incompetency leveled against him, charges which still had not been—and possibly could never be—proven, charges that he could never acknowledge as true. It is easy to understand why Mulholland subscribed to the sabotage theory for the dam's failure, but it was impossible for him to accept that the destruction of his own engineering achievements may have been instigated by his own actions.

Like mythic heroes of antiquity, the perpetuated image of William Mulholland, the great self-made engineer, conqueror of desert sand and rock, deliverer of the life-giving water, the vigorous, infallible leader beloved by his men, was at the time of the tragedy an integral part of the psyche of the city of Los Angeles, as important and necessary to its esteem as the great aqueduct itself. And perhaps nobody in Los Angeles bought into the myth of William Mulholland more than William Mulholland himself.

"I have a job to do. And damn it, I will very well do it. And if I have to step on a few bruised egos or hurt a few so-called educated types, I will. But by God I made a commitment to the people of Los Angeles and so help me God, I intend to finish the job," Mulholland told Van Norman one day in the blistering heat of the Mojave Desert, venting his exasperation with the political wrangling over the aqueduct.

The inquest was much more than a legal proceeding to establish cause of death. It was, also, the trial of William Mulholland and no one knew it more than he. Few individuals in American history have risen to the heights of fame and achievement that Mulholland had secured, only to suffer such extreme public disgrace and personal sorrow. The tragedy of the St. Francis afforded the Chief the occasion to rise to his highest point of inner strength. Mulholland's immediate and unwavering acceptance of the responsibility for the disaster would earn him the admiration and sympathy of nearly every reasonable person, save those perhaps who had lost a loved one in the catastrophe. During the inquest, and following, letters and editorials running side by side with those which angrily attacked him filled the newspapers, assuring Mulholland he would not be abandoned in his darkest hour by the city he inspired.

Ironically, the year the dam collapsed was the same year Mulholland succeeded, after three long years of lobbying in Sacramento, to obtain the California state legislature's approval of a bill creating the Metropolitan Water District. After that, Los Angeles's aggressive annexation program of Owens Valley land and water was halted.

But for now, as the inquiry reconvened, Mulholland's alleged shortcomings as an engineer were working to undermine his diplomatic achievement in preserving his position as the head of the city's water program through more than forty years of political flux. The future of the Swing-Johnson bill, authorizing construction of Boulder Dam, was still in question. The violent history of the Owens Valley water wars coupled with the St. Francis disaster had caused the city's Congressional supporters a degree of embarrassment they could ill afford in the midst of their delicate negotiations over the Boulder Canyon project. As historian William Kahrl pointed out, by now, as "the architect of both the dam and the city's policies toward the Owens Valley ranchers, Mulholland was a liability that could no longer be sustained."

AS THE WITNESSES and audience settled into their seats, Asa Keyes wisely chose to continue his attack on leaks and departmental neglect. Harvey Van Norman was the first to be called. When questioned about the selection of the site of the St. Francis, Van Norman, though deeply bereaved by the catastrophe, was not so subdued in his replies as was the grief-stricken Mulholland. He sharply retorted that the dam's site was sound and its location justified. "The dam wings were joined soundly to the hills, on both the east and west, on a foundation of very hard shale," he said tersely. When asked by what other geological name the shale was called, Van Norman said it was called "schist."

Keyes was ready for the anticipated answer and pounced on it like a cat on a mouse. "As a matter of fact, schist is really nothing more or less than 'rotten rock', isn't that so?" asked Keyes, referring with usual histrionics to a field geologist's term for porous rock, or conglomerate, which percolating water could gradually reduce to a spongy precarious mass.

Although he preferred calling it shale, "schist," or "rotten rock" was safe, affirmed Van Norman, readily admitting three other dams—the Barret in San Diego, the Arroyo Seco in the San Fernando Valley, the Mulholland Dam in the hills above Hollywood—were all built by Mulholland on the same formations and were in fact "near duplicates" of the failed St. Francis. Keyes could not have been more pleased at the news, and Van Norman's "rotten rock" testimony sparked a heated exchange over the safety of the dams. By noon the next day, hundreds of angry citizens carrying placards were marching around the perimeter of the Mulholland Dam demanding it to be emptied.

Reacting to the ensuing headline that blared, "Film City in Fear of Dam Breaking," and charging that his life and property were in imminent danger, David Horsley, a motion-picture producer and Hollywood businessman, filed a multimillion-dollar suit in Superior Court against the City of Los Angeles and the Water Department. According to Horsley's complaint, 250,000 lives and $300 million in property would be threatened if the dam were to break. "The people of Hollywood are justified in believing that like the St. Francis, the Hollywood Dam is doomed to failure—the dams were built by the same men with the same materials and by the same plans," argued Horsley.

To Keyes's delight and to Mulholland's mortification, in response to the public panic, and without consulting Mulholland, the Water Department emptied the reservoir by 25 percent.

W. B. Mathews, special counsel for the department, sought a general demurrer with little fanfare; this was in stark contrast to the old days of the aqueduct when he and Mulholland were constantly in court battling special-interest groups. The Chief's keen Irish wit would often turn an opponent's attack into a striking point for Mathews, and the two men together rarely lost a case. Now, squelched by the momentous events closing in around him, the Chief sat in morose silence in the hot, humid inquiry room listening to his second-in-command struggle to his defense, affirming his choice of the site as sound.

The issue of leaks, and knowledge by the department of those leaks, continued to be the focus of Keyes's attack. Even through Keyes produced reports that the flow of the Santa Clara River several miles below the dam had increased over ten feet by 7:00 A.M. the day before the break, Van Norman was adamant that the leaks he and Mulholland had inspected were crystal clear and normal.

"I want to bring out my intimate relationship with Tony," Van

Norman said firmly, referring to dam keeper Tony Harnischfeger. "I had known him fifteen years and we had the utmost confidence in each other. If he had been alarmed over the leaks he would have told me." Backing up Van Norman's statement, Dean Keagy, a shipping clerk at Powerhouse Number One above the dam and a worker on the dam during its construction, and also the aqueduct, testified that he had noticed nothing unusual when he drove past the dam at 11:30 P.M. Monday. Not to be deterred, Keyes quickly redirected his questions to the quality of the concrete and other materials used in the dam's construction, asking Keagy if he had seen any red clay or dirt go into the concrete mixing machines, prompting Keagy, as quickly, to respond that he certainly had not.

But others claimed they did. Knowing this time he was on safe ground, Keyes again posed the question that had sent him into throes of panic earlier. "Do you know of anything that would offer us here in this room a clue as to what caused the failure of the dam?" queried Keyes of Richard Bennett, who had operated a concrete-mixing machine during the construction. Aside from the rumors he had heard regarding old iron bed posts used as foundation girders, the florid, boozy-faced Bennett, attempting to substantiate Keyes's contention that clay had been mixed into the concrete as a cost savings, said he "saw things go into the mixer that shouldn't have gone in." The clay, or red mica schist, he added, was a common element of the terrain from the San Francisquito canyon.

The next morning, the fourth day of testimony, Keyes performed a dramatic experiment for the benefit of the jury. He held up what appeared to be a small ordinary piece of rock and identified it as coming from the ill-fated dam. Then he plopped it into a glass of water, waited three minutes and stirred the glass with a lead pencil. The rock, about the size of a walnut, disintegrated, coloring the water a muddy red and leaving a thick layer of sediment on the bottom of the glass. Without comment, but with eyes gloating, he set the glass down in front of the jury.

Watching the display was an excruciating moment for Mulholland and Van Norman who feared the unscientific experiment would inflame an already angry public and could sway the minds of the jurors. The so-called experiment was indeed sensationally played up in the newspapers causing further outcries for Mulholland's head, and, in a morbid twist, became for a time a popular parlor game. To play, one merely had to drive up to the dam site, gouge out a sample of the clay, take it home, place it in a glass of water, and then wager on the time it took to dissolve.

To confirm the results of the clay experiment, Keyes called upon David Matthews, a dam laborer who gave sensational testimony to coroner Nance that the dam was not safe. Matthews worked at Powerhouse Number Two along with Ray Rising. He stated he had been extremely fearful about the dam, fearful of the "five-mile lake that would go crashing through the canyon if the dam ever broke." He added that the west hill of the dam had become "soft and boggy" as a result of the continual day-by-day seepage.

Matthews claimed that the leaks increased daily over the two-week period preceding the collapse, and on Saturday, two days before the flood, the entire west hill was saturated and looked like it was ready to give way at any moment. The day before the break, Matthews continued, water tinged with red clay oozed from the hill under the west wing. Walter Berry, Matthews's boss, had ordered him to stuff logs into the aqueduct drain pipes to relieve the pressure and seal them with oakum. Berry then left to report to Tony Harnischfeger. "When Berry came back, he had some pretty sensational news."

"And what was that?" asked Nance.

"Well, he called me aside and said, 'This is confidential—but the dam's not safe. I got orders from Tony to put logs in all the pipes.'" Going off duty that evening, Matthews said he passed Harry Carey's ranch on his way home and was worrying about what Berry had told him when he saw his brother who lived in the area coming in his direction in his car. "I felt I owed a duty to him, being my brother and all, and I thought I'd ought to tell him. I yelled at him—For God's sake, get your family and get them out of the canyon, and I told him about the dam. He said, 'Dave, I'll move to

Newhall tomorrow.' That tomorrow never came. That night she broke." At this, Matthews buried his face in his hands and wept.

A short recess was ordered, the inquest resumed with the testimony of Chester Smith, a working rancher who had shared his house with the Nichols family in the San Francisquito canyon.

"Water began to seep from the west wing a month ago and I told him, Tony, about it," said Smith. "Tony said he wasn't worried, but that if the dam did break it would be just at the west wing. I went up to feed my stock above the dam the day before she broke. I met Jack Ely and asked him if he was going to flood us out because the west bank was saturated with water and water was slopping over the top of the dam. I know he was only joshing, just joshing, but Ely hollered back—'We expect it to go out any minute!' Ely was working at the dam and I told him he'd better turn some of that water out. Why, it was splashing over the top and the leak was bigger than ever.

"But I was afraid the dam might fail and I slept out in the barn that night and kept the doors open. The dogs woke me up at 'bout midnight. They were barking and just raising the dickens. I knew something was wrong when they acted like that. Then I heard that awful noise—saw trees and telegraph poles breaking and electricity flashing from the power lines. I knew what was coming. I'd been in floods before. I ran out in my nightshirt and yelled to Nichols in the house. 'Dam's broke!' I shouted, loud as I could. I had to drag them both out of the bed and we all started running up the hill. We had to pull Mrs. Nichols along, Nichols and I. We got out alright with the water all around us. It was 125 feet deep when it tore through my ranch. Stripped it clean as a whistle.

"When the water hit the wires it made lightening, but that soon stopped and we couldn't see anything, but could hear the roar. When we reached safety on the hill the water had covered everything. My feet were bleeding and most of the nightclothes I had on had been torn off. We stayed up there above the water till dawn, till it went away."

Keyes presented more witnesses who had personal knowledge of the leaks. Mrs. Anna Scott, resident of Mint Canyon, testified that she was driving up San Francisquito canyon in her new Ford on Monday, March 12, but had turned back when she noticed a large volume of muddy water flowing down the stream bed and threatening the bridge.

Robert Atmore, a rancher at Hughes Lake several miles above the dam, testified that he had noticed a bad leak on the east wing of the dam. "It occurred to me more than once it was dangerous because I had seen it before. I talked to my friend Harry Burns who was living below the dam about it and told him he ought to get out of the canyon—that if it ever broke, he'd never get out alive. He just laughed. He died that night."

After presenting still more testimony from witnesses regarding the leakage and rumors of cover-ups by dam employees to substantiate his theory of criminal neglect, Keyes held a dramatic press conference in his office that night. "Should the committee's investigation show evidence of knowledge of the defective construction, I will ask the Grand Jury to return an indictment charging murder. Should evidence propose either negligence in the construction or conscious knowledge of danger to lives through such construction, indictments charging voluntary manslaughter will be asked."

Keyes knew that by week's end public sentiment and cold, hard statistics would be working to his advantage. He was right. The toll would stand at 511 dead or missing, with 381 bodies recovered, 297 bodies identified, and 74 corpses still unclaimed. Property damage would exceed $20 million.

16

JUDGMENT

We look for judgment,
but there is none.
Isa. 59:11

During the two-week period of the inquest, Mulholland drew further into himself. Picking listlessly at his dinner, he asked Rose: "What's the matter with me? I see things, but they don't interest me anymore. My zest for living is gone." Rose knew her father's plaintive question was rhetorical, but he was, of course, correct in his self-diagnosis. The irrepressible spirit that had characterized him since youth was gone, washed away with the collapse of the dam. Returning home at night from the inquest, he would pull the blinds, lock the doors, and sit most of the night brooding in his study, unable to concentrate on reading or listening to the classical music he loved. When he tried to sleep he would toss in bed and wind up pacing the house. From her bedroom Rose could hear his footsteps "creaking the house in pain" as they padded across the oak flooring.

Fearing their presence would sink him even deeper into depression, loyal friends and colleagues declined to visit. For some, unable themselves to face the dark despair inflicting their once-vibrant Chief, this reason served as a pretext for not calling.

Many of the physical traits that were noted at the inquest by reporters, including his masklike face and speech difficulties (dysarthria), alternating tremors, and stooped posture, were attributed to the nervous system disorder of Parkinson's disease, but the deeper, less obvious symptoms of depression and acute anxiety that had enveloped him was described to Van Norman by Dr. Raymond Taylor as "plain old shell shock."

Anxiety-neurosis had been observed in soldiers in World War I, but once a soldier was removed from the stress of combat, symptoms generally disappeared. For Mulholland, however, still in the heat of battle, the tragedy proved to be a continuing source of anxiety. As tension increased, Mulholland found himself not only unable to rest or concentrate, but also experiencing heart palpitations, severe migraines, weakness of the limbs, and heightened irritability. Confiding in Van Norman, who related the information to Taylor, he also admitted to feelings of uneasiness, fear, dread, and panic.

In the immediate aftermath of the flood, Mulholland largely directed his energies to the immense details of the rescue operations, disassociating himself from his enormous sense of personal responsibility for the tragedy. He told Taylor that the horrific scenes of destruction during those first days had seemed unreal and remote, a pyschological defense not uncommon in persons confronted with unexpected tragedy. But on those rare nights when he was able to drop off to sleep, he would suddenly awaken in a cold sweat from the nightmare that had been haunting him since the disaster: In the pitch black night, with lightning flashing ferociously above and raging waters ripping away huge chunks of concrete below, the great remaining center piece of the dam loomed up in his dreams as a giant tombstone, names of the flood victims etched on it in a never-ending list.

In reality, the last standing remnant of the dam was to be buried in a five-ton blast by the Department of Water and Power not long after the disaster. Amid a shower of debris and a rain of stone, the huge derelict, over ninety million pounds of concrete, came tumbling down, forever eradicating the last tangible image and hideous eyesore of the once-magnificent St. Francis Dam.

During the final days of the inquest, Mulholland was the first man seated in the Coroner's courtroom. He would wave his hand somberly to the scores of reporters and photographers upon arrival; then sitting quietly, he would begin each day by absorbing himself in the details of the interrogation. But by day's end, his melancholy gaze would drift to his hands or to the scenes of life out the window.

On the morning of March 23, after the dramatic testimonies of laborers Bennett and Matthews, and under strict instructions by coroner Nance to discuss the case with no one but themselves, the jurors traveled by bus to the dam site for a firsthand look. Twenty-four hours earlier, a heavy rain had saturated the ground, activating the pungent smell of the decomposed matter, causing some jurors to clasp handkerchiefs to their noses as they slogged through the ankle-deep mud to the center piece. There they looked out at the valley below and saw the devastation wrought by the "tide of doom," as the *Los Angeles Times* had called it, then scrambled down to the floor of the canyon and along the adjacent hillsides, pecking here and there with miner's picks and hammers, taking samples of rock and concrete fragments to be analyzed in Los Angeles later. The staccato "rap-rap" of their small hammers and the echo of their muted conversations could be overheard by the still-patrolling guards and the ever-present press corps restricted to observing them from a nearby ridge.

First, the jurors obtained samples of the questionable shale or "schist" at what was the west wing of the dam, stuffing them into the pockets of their baggy, city-issue coveralls, and then cautiously moved across the precarious, mud-caked ridge that had supported the dam's foundations and around the great center piece to the crest of the east wall, and there took more samples. The inspection lasted four hours, and during the ride back to Los Angeles the jurors viewed for the last time the path of destruction as it had roared through the valley.

Back in Los Angeles, using the latest core-cutting devices and modern scientific instruments to test crushing strength, it was ascertained that the quality of the concrete could be eliminated as a cause of the failure. The concrete was "clean, of good quality and adequate for the job," stated the consensus report of the jurors, adding that after sixteen years, Mulholland's aqueduct was still standing and functioning with precision, and he had used the same concrete mix in the construction of the St. Francis. Having settled the concrete question, the jurors turned to the problem of determining which side of the dam fell first.

With the absence of any evidence or testimony corroborating the dynamite theory that both sides were blasted simultaneously, the jurors were split into two camps—those believing the initial break occurred on the dam's western side and those believing it occurred on or near an old earthquake fault line running on the east side. Reconstructing the events, the pro-west jurors speculated the water working against the soft, quartz mica schist on the west canyon wall finally eroded its support. Once the west wing went, the east wing, anchored into much denser material, collapsed in three places, toppling slightly inward as the escaping water undermined the foundation weighing on the canyon floor. Concrete dam fragments from the west wing had been discovered far downstream in the canyon, whereas huge concrete slabs from the east wing were piled in a heap close to the remaining center piece. Those who believed the east side fell first speculated unseen water percolating under its foundation resting on the old fault line caused the same action to occur.

Both theories were flawed. A Stevens gauge, an instrument that measured water flow into and out of the dam, was recovered from the control tower located in the center piece immediately after the break. It indicated the reservoir had lost as much as twenty-two hundred cubic feet per second in the time frame just before the failure. However, since the gauge—later seized from the Department of Water and Power by the district attorney's office—could not accurately pinpoint the exact time of the loss, it could be used by seepage or sabotage theorists alike.

If the gauge's water-loss readings were accurate, then leakage into the canyon prior to the dam's failure would have been enormous, warning dam attendants on duty of impending danger. However, dam employees like Ray Rising had testified that they saw nothing unusual to arouse suspicion prior to the dam's collapse. One of the dam keepers was seen on the parapet of the dam calmly smoking a cigarette only one hour before the failure. He had given no indication of alarm. The dam keepers had all perished in the flood, and as far as was known, there were no living witnesses of the dam's collapse.

Additional conflicting testimony further baffled jurors. Many witnesses said they had lost faith in the dam's safety, but admitted that they did not consider themselves in such immediate jeopardy to warrant leaving the area. Too, they did not voice their concerns to water department employees or to the dam's engineers, though these officials were readily available.

Quick-thinking Keyes recognized that the contradictory testimony could work to his advantage. Credible testimony about seepage locations would tend to support his west-wing collapse theory. As long as he directed the attention of the jurors to a west-wing versus east-wing failure, the "dynamite factor" could be safely ignored.

To add to the jurors' confusion, E. H. Thomas, a rancher asleep at his home one mile below the dam on the night of the break, told jurors that an earthquake may have caused the break. Thomas's mother, an insomniac, awakened him when the electricity suddenly went out at about 12:05 A.M. The house was trembling on its foundation and continued to shake for almost ten minutes. Thomas slipped on his trousers and went outside to see what was going on. He was greeted by a loud roar and looked toward the direction of the dam to see the black head of the water "ninety-nine feet high above the bed of the San Francisquito Canyon sweeping away every damn thing in its path."

One wild-eyed would-be witness, who had to be restrained by

sheriff's deputies, broke into the jury room shouting a garbled theory that the Ides of March had caused the catastrophe, hoping to give credence to Mulholland's widely reported hoodoo testimony.

And, momentarily, Asa Keyes was unsuccessful in keeping the word "dynamite" outside the courtroom. Called to testify on leakage, gruff-mannered maintenance engineer J. H. Bouey shocked Keyes and the court when he stated matter-of-factly that he personally knew 625 pounds of high-powered dynamite had been intentionally set off by Tony Harnischfeger three hundred feet below the dam's west wing on the Tuesday preceding the break to start grading a new road behind the western dike wall. However, Bouey went on to explain that no new leaks had sprung, and there had been no unexpected earth or rock movement due to the blasting. "As far as I'm concerned, sir, there was never any indication to cause fear that the dam would go out."

Keyes attempted to get Bouey to admit that it was severe leakage that caused the need for the road work in the first place, that the road had become so saturated by continuous seepage as to be in bad shape.

Privately Van Norman told Mulholland that the leakage testimonies of Keyes's "would-be engineers wouldn't hold water," the unconscious pun drawing a chortle from the usually somber-faced Mulholland. "The ones who stated they knew the dam would burst are liars or criminals," Van Norman continued, outraged. "If they're lying they should be muzzled. If they actually knew that the dam was weak or that it was constructed negligently, they're criminals because they failed to tell the proper authorities. The parties spreading this propaganda should be made to answer."

ALTHOUGH MULHOLLAND LISTENED keenly to the bolstering logic of his young ally and agreed with it, it offered little in way of consolation. While the jury debated the varying issues, news reports continued to surface about the misery of the flood victims. A gaunt, eighty-two-year-old man caught in the flood wandered dazed and disoriented into the town of Hanford, miles north of the Santa Clara Valley. Bewildered, the man said he had been wandering since the dam went out fourteen days ago. He was unaware that his wife, also caught in the floodwaters, had drowned. That same evening, the bodies of two more victims, both male, were washed up by the tides on San Diego beaches, 120 miles south of Los Angeles.

The flood had scarcely spent itself into the Pacific Ocean when scores of fast-talking lawyers swooped into the Santa Clara Valley to sign up grieving and angry clients on a contingency basis, promising high-dollar negligence claims against the city. Branding them vultures, Ventura District Attorney James Hollinsworth executed disbarment proceedings against any attorneys deemed engaged in ambulance-chasing in the flood-stricken zone.

The question of repaying the huge losses became a critical and sensitive issue for city leaders and the Board of Control. The multimillion-dollar disaster could not have occurred at a more inappropriate time, in that the Boulder Dam bill now before Congress was critical to the commercial interests of Los Angeles. Recognizing immediately that a spirit of cooperation and fair play would serve everyone's best interest, and in the end prove to be less expensive than unending legal wrangling and litigation, the Chamber of Commerce and the edgy Board of Water and Power Commissioners pressed through local newspapers for the city to pay immediate out-of-court reparations.

Mayor Cryer was an early advocate of the idea, despite argument from City Attorney Jess Stevens that until legal liability was fixed, no payments could be made. Stevens maintained the city should be liable for full damage only if carelessness or neglect were established, factors, of course, which were now before the jurors of the coroner's inquest. "If this proves to be what is called an 'act of God,' or if the mishap was beyond human responsibility, the city cannot be liable."

Despite Stevens's position, the City Council released $1.5 million to begin restoration. To any dissenters, the costly Owens Valley water wars only had to be mentioned. From a public relations viewpoint, Los Angeles could not afford another long, protracted battle with angry valley residents. "The city will take care of all property damage," Cryer stated firmly.

Despite Mulholland's fall from grace, he was still the Chief of the city's Department of Water and Power and cleanup of the Santa Clara Valley was the department's number-one priority. Men, mule teams, and machinery were assembled along a thirty-mile front from Santa Paula to Piru as the month-long, million-dollar reconstruction and mop-up program began. A thousand laborers and skilled contractors from Los Angeles were recruited to initiate the unprecedented, fast-paced campaign. Under the aegis of the department, Van Norman was selected to head up the efforts.

Seeing firsthand the total scope of the damage, Van Norman was appalled. But later, telephoning Mulholland and boosting his spirits, he said how surprised he was that much of the valley had been spared the flood's wrath. Van Norman, as opposed to the taciturn Mulholland, was the eternal optimist. Only one-twentieth, about ten thousand acres, of the cultivated region from Castaic to Saticoy was touched by the flood. Bungalows and cabins near the river were obliterated, but the business and residential sections, except for the community of Santa Paula, had completely escaped. Van Norman told a relieved Mulholland he believed it was a miracle.

Nonetheless, much laborious and heartbreaking work lay ahead. There were countless heaps of debris spread across the devastated area that had to be disposed of, adding to the severe unsanitary conditions; and the hundreds of tractors sent by the city of Los Angeles were working around the clock removing the masses of wreckage.

Grieving Santa Paula residents feared the debris might contain bodies of missing victims, and when the charred body of a baby was found after one pile of debris had been set on fire by mop-up workers, they went wild. To calm their concerns, Van Norman supervised teams of shell dredgers who vigilantly monitored the tractors. One man was stationed at each dredger to make sure no more bodies were uncovered before disposing of the debris. Only 297 of the missing bodies had been recovered and many more were feared buried in the debris. Ventura Coroner Olive Reardon issued a warning to those handling the bodies to wear rubber gloves and protective clothing to prevent the danger of infection.

Every care was taken to make positive identification of each corpse. Photographs were taken and distributed for family members to view, jewelry, clothing, and teeth were examined for any evidence as to a deceased's identity. There were so many unidentifiable corpses recovered that the coroner and department officials had to use 3 × 5 cards inscribed with notes to identify each cadaver. Phrases like "brunette," "cesarean section," "bald," "missing teeth," were scratched on the cards in hopes that it could aid in the identification process. As the number of cards grew, Van Norman and Reardon had the painful duty of making the match. Death certificates normally sent to the next of kin were stapled on the cards of unclaimed bodies, as there were no known individuals to whom the certificates could be sent.

The mangled body of a little girl was found beneath one of the piles of debris on what used to be the Stark Ranch. Her height was three feet, ten inches, and her hair was dark brown. Such was the force of the flood that her teeth were cleaved to the gums, and her naked body was clothed only by one shoe with the laces untied. The sight of the little girl affected Van Norman, who had seven brothers and sisters, but was himself childless, more than any of the other bodies recovered.

True to Cryer's word, the city took full responsibility for the catastrophe and quickly paid most of the claims without contest. In all, over 1,000 claims were filed, including 336 death claims totaling $4,864,006.77. Nearly $1 million was paid for real property claims, and another $1.7 million was paid in land reparations. Hundreds of homes and businesses in the flood district were rebuilt, utilities repaired, county buildings and services restored, orchards replanted, and ranch lands rehabilitated. Eventually, cleanup of the Santa Clara Valley cost the city of Los Angeles over $7 million. For far less, Mulholland could have bought the Long Valley site from Fred Eaton, even at Eaton's highest price, and built the critical storage reservoir as well, a thought that only added to Mulholland's grief.

The swift resolution of money claims and Van Norman's conscientious direction of the restoration avoided further embarrassment to the city, protected the Department of Water and Power from any severe recriminations from the Santa Clara Valley, and safeguarded the essential passage of the Colorado River Plan.

THE INQUEST, now in its next-to-last day, had all of Los Angeles anticipating a final swing of executioner Asa Keyes's sword that, as he had boasted weeks before, would deliver them Mulholland's head.

To establish Mulholland's responsibility as chief engineer of the Department of Water and Power and to substantiate the popular criticism that he was a dictator in the water affairs for the city, heeding no one's advice but his own, Van Norman was again called to the stand. Asked who had the final decision on selecting the site, Van Norman admitted that it was Mulholland's sole decision, although he had discussed it in conferences with other engineers in the department.

"And of course you and these engineers voiced your own opinions on the selection of the site in these conferences?" Keyes asked.

"Well, Mr. Mulholland holds a unique position in the engineering world and I think his opinions on such matters are as good as can be obtained," replied Van Norman firmly, adding that Mulholland often changed plans in accordance with suggestions from his staff.

"But always reserved the major decisions for himself," interjected Keyes.

"Yes."

Pressed by Keyes, Van Norman admitted that Chief Engineer Mulholland held a position of responsibility to the city "that no man should have shouldered." His refusal to shift responsibility to others was characteristic of the man known during his more than fifty years in the water department as a "straightforward fighter."

Invariably, the brilliant, self-taught engineer was light-years ahead of his subordinates in solving even the smallest problems and, wanted or not, responsibility was thrust upon his shoulders. Unlike bureaucrats who protected their empires by maintaining the status quo, Mulholland made a point to fill his staff with bright young administrators and engineers, making it his business to nurture their careers. In return he got back unyielding loyalty all the way down to the lower ranks of the department. But, as any new eager employee was quick to discover, the Chief always had the final word. All decisions, major or minor, in the operation of the department and the construction of the dam were his. Only when it came to paperwork did Mulholland relinquish control. Loathing sitting at his desk, preferring to work in the field, he relied on others, especially Van Norman, to handle the administrative nuisances.

Like a matador waiting patiently in the center ring for his picadors to weaken the already dazed, exhausted bull for the kill, Keyes next allowed the jurors their first opportunity to question Mulholland when he was called to the stand. Repeatedly the jurors asked who had actually done the work in designing the dam.

"Our engineering force collaborated on it," Mulholland said wearily, adding that state engineer John Hendrix had inspected it at completion even though a state inspection was not required for municipal dams. Mulholland said he requested the inspection as added proof of its safety. "Mr. Hendrix spent half a day in his inspection, which is an unusually long time. I was confident the dam was a fine, safe dam."

Mulholland informed the jurors that over two dozen test holes were drilled into the conglomerate rock foundation on the canyon sides prior to construction. "I don't know why these other engineers testifying here didn't mention that. They must have known about them. The holes are there yet. I can show them to you. We filled them with water, left the water there two weeks and then we had to bail it out. It had not soaked in."

One skeptical juror produced a chunk of the reddish schist taken from the dam site, and said that it was a fragment from a piece that disintegrated when he had played the parlor game at home. "Well,"

Mulholland commented, shrugging his shoulders. "That sort of thing was uncommon, but the whole foundation wasn't made of that. You can find almost any mineralogical specimen you want to look for up there," adding that it meant nothing.

With patience and straightforwardness, Mulholland continued to answer the juror's questions, but when the rumor was brought up by Keyes that Mulholland had been letting up in his rigorous hands-on style of supervision by delegating more duties to subordinates, Mulholland took issue to the implication that his physical and mental faculties were eroded by age and had affected his decisions in the construction of the St. Francis.

"I haven't been letting up," he protested. "I've been working harder than I ever did in my life. I haven't had a day off—the only vacation in my lifetime, I took about three months ago, through the Panama Canal to New York. In regards to my men, I am the first up in the morning, and the last to go to bed. There are a very few that beat me in the office in the morning." Then he concluded in a tone of despair that gave sympathetic pause to just about everyone in the room, except Asa Keyes, "As far as letting up is concerned, I wish I could. I believe I will have to very shortly, this thing has got away with me.

"Don't blame anyone else," he continued. "Whatever fault there was in the job, put it on me. If there were any errors in judgment—and it's human to make mistakes—the error was mine.

"I appreciate the job you have before you," he said, facing the jury. "I haven't anything to conceal. I'm waiting to hear the report of the geologists and, whatever it is, I hope it will be published to the world."

The crowded courtroom listening to the aging engineer sat in silence. "I am giving you all I know. I swear to God on my oath that I am."

KEYES'S CASE against Mulholland came to its climax the following day, Thursday, March 29, when a special board of investigative geologists and engineers were called to the stand. Hired by the district attorney's office, they offered their findings on the collapse in a 112-page confidential report hand-delivered to Keyes. Keyes chose the final day of the inquest to dramatically introduce it and the geologists into the proceedings.

Key testimony was presented by Allen E. Sedgwick, a congenial, highly regarded professor of geology at the University of Southern California.

Sedgwick declared, as did the other geologists, that the eastern hill of the dam site was composed of a vein of quartz-mica schist, which continued across the stream to the west side, up the bank, running into a much softer conglomerate formation than that on the east side. The conglomerate was so badly weathered or "rotten" it would absorb great quantities of water and could not sustain great weight loads. In his opinion, it was evident that water had oozed through the entire conglomerate formation on which the largest portion of the dam was constructed. When the saturation was complete, the entire west side ground mass gave way under the immense pressure of the water in the reservoir, tearing a hole sixteen feet deep and forty-two feet wide into the hillside.

"The dam then failed because of poor foundation, is that what you're saying, Dr. Sedgwick?" asked Keyes somberly as he led the geologist through the series of questions that he knew would drive the last nail into Mulholland's coffin.

"The failure was due to defective foundation material, some of which, while reasonably hard when dry, becomes soft and yielding when saturated with water."

"And how could this determination have been made?"

"Any competent geologist, once he studied the location, would have recognized that the dam was situated on incompetent geological formations."

"Do you know if any geologist ever studied this site for defects?"

"No, sir, not to my knowledge. The site selection was completely in the hands of the Chief of the Department of Water and Power."

"Should a dam have ever been built at this location?"

"No, sir. The dam as designed should not have been constructed at this location."

"Was it feasible to erect a safe dam at this location, Dr. Sedgwick?"

"No, sir. It was not."

After the testimonies of the geologists, for all intents and purposes the coroner's inquest was over. Only the verdict remained.

A GROUP OF Mulholland's closest friends and admirers decided to put together a private candlelight banquet at the Los Angeles City Club in tribute to their Chief, to bolster his spirits. Coaxed by his family and Van Norman, Mulholland attended, accompanied by Rose.

Toastmaster for the occasion was C. A. Dykstra, who raised his glass and lauded Mulholland for his distinguished services to the city of Los Angeles. "A man with a mind remarkable for its breadth and brilliant wit. A man who can build an aqueduct, and a man who can also, beside a mountain campfire, while he boils his trout, discourse on profound structural geology. A man whose life has been spent in public service for the benefit of the masses in the land of his adoption. Remarkable for his originality of thought and analysis, yet equally active in the practical application of these ideals. Original in the minute details of construction, yet brave to the limit in conceiving and assuming the responsibility of the greatest projects. Kind, generous and true to the public welfare, he stands as an example of what the applied scientist can do for his state when he holds his brief for the people." The City Council, Mayor Cryer, the Governor of California, friends, colleagues, and fellow workers rose to their feet, clapping in a spontaneous and sincere standing ovation. Accepting their applause in a brief, typically humble, "peppered" speech, none but daughter Rose and a handful of closest friends knew the private torture he endured during the inquest and continued to suffer.

DOZENS OF WITNESSES, engineers, amateur geologists, and survivors of the midnight flood had been heard—and yet no clear consensus about what caused the tragedy had emerged. The responsibility of determining the cause of the St. Francis disaster was now in the hands of the coroner's jury, but not before Asa Keyes had delivered a short but powerful closing statement.

"Ignoring warnings regarding the possible instability of the site, Chief William Mulholland may have succumbed to the worst traits of the autodidact—self-absorption edging into arrogance, a hostility to experts approaching culpable negligence.... Los Angeles had granted Mulholland the very power that ended in the great tragedy of the St. Francis.... In selecting the dam site Mulholland was acting alone, and it is up to you, Gentlemen, to determine who, if anyone is guilty of causing this catastrophe."

Keyes also reviewed the law's provisions as to manslaughter, elaborating that "an honest mistake is not criminal negligence. It would be monstrous to charge a man with murder or manslaughter because of an error in judgment. If, in the erection, inspection, and maintenance of the St. Francis Dam ... ordinary care, prudence, and honesty were exercised to do what was charged upon the builders—no criminal action could lie as a result of the dam's failure.

"It has been declared that if geologists had been employed by the Department of Water and Power, they might have disclosed the treacherous foundation on which the dam was built. They were not so employed. You must determine if it was criminal not to employ them. The matter was left to the judgment of one man and he may be the only man alive who can know for sure.... If the builders of the dam used all the care and prudence of which they were capable, if they acted according to their best judgment, it is not a matter for criminal prosecution."

Coroner Nance then addressed the jury: "Members of the jury, you are instructed that testimony based upon theory or suspicion only is not the best evidence, and that you are to give such testimony only such consideration as ordinary prudence would dictate. It becomes your duty to determine how, when, and where Julia Rising came to her death, as well as all the others whose names have been read into the record of this case, so that we can prevent this horrible tragedy from ever occurring again.

"Gentlemen, God speed you in your deliberations."

17

REDEMPTION

Redeem us
for thy mercies' sake.
Ps. 44:26

WHEN ASA KEYES announced his intention to indict William Mulholland for manslaughter, he was bitterly accused of "over-reaching in a vicious abuse of official power" by prominent judges, lawyers and city officials. He had been denounced in the past by colleagues and judicial officials for engaging in shoddy and petty prosecutions, and brusque reprimands for his lack of professional standards were not uncommon. His performance in the inquiry was not out of character. Compared to "a buzzard of the human species that forced itself into the public eye by condemnation and vilification of innocent public men in an attempt to appear like a hero vigilantly defending the people's interests," Keyes himself was vilified in withering, scornful public denouncements and asked to resign.

As the *Los Angeles Times* commented gravely: "The trouble is that Mr. Keyes is not, in fact, the directing head of the District Attorney's office; the office has, in fact, no responsible head.... The deputies have been allowed to run wild, most of them are careless and incompetent and some may be corrupt. If Keyes cannot or will not attend to the duties of his office, he should get out, and if he does not get out, he should be compelled to do so."

Many Los Angeles citizens protested that Keyes had gone too far in his attacks against Mulholland, and that Keyes's verbal abuse of the Chief during the coroner's inquest was cruel and unnecessary. Keyes was charged with "official blundering and dereliction" by two Superior Court judges, and the words of one editorial writer reflected the city's anguish over the fate of their beloved hero and embarrassment at the cruel harassment by the likes of Asa Keyes. "The people feel a kindling sympathy for Chief William Mulholland ... remembering the fable of the lion who was tormented by the gnat." The tragedy of the St. Francis Dam had brought death and destruction to the Santa Clara Valley, and now had delivered "evil politics" to the city of Los Angeles as well. Like many public prosecutors voted into office espousing moral indignation, Keyes would, in his rampant, often suspect zeal, wind up an embarrassment to the very establishment that got him elected.

But for now, as the entire city anticipated the jury's verdict and while Mulholland tortuously waited, ensconced with Van Norman in his office at the department or at home with Rose, Asa Keyes basked in the spotlight, speaking to reporters at the slightest pretext and dining with colleagues in expensive restaurants near the Hall of Justice, dismissing any criticism of his tactics as poppycock. "The fact is ... to my supporters, I am a hero vigilantly defending the people's interests," Keyes righteously told reporters.

On Friday morning, April 13, 1928, two weeks after the inquiry adjourned and one month after the collapse of the St. Francis dam, the nine jurors concluded their deliberations and filed into the inquest room, pushing their way through the throng of newsmen and onlookers waiting outside.

The men sat with backs straight, hands folded, expressionless, as Juror Lawrence G. Holabird read aloud their decision in his high monotone voice.

> Responsibility for the St. Francis Dam disaster is placed on the Bureau of Water Works and Supply and the Chief Engineer thereof, and those to whom the Chief Engineer is subservient, including the Department of Water and Power Commissioners, the legislative bodies of city and State and the public at large. The destruction of this dam was caused by the failure of the rock formations upon which it was built, and not by any error in the design of the dam itself or defect in the materials on which the dam is constructed. The gravity section accorded with standard practice, and would have produced a safe structure if it had been built upon hard, impervious rock, as was supposed to be the case by those who built it.
>
> On account of the great destruction wrought by the disaster and the absence of living eyewitnesses, much important evidence bearing on the cause of the failure was obliterated, making it impossible to determine, with anything approaching complete accuracy, the exact cause of the initial break and the sequence of events thereafter.
>
> It was apparent that the entire personnel of the water department had an unusual degree of confidence in Chief Engineer William Mulholland and relied entirely upon his ability, experience, and infallibility in matters of engineering judgment. However, Chief William Mulholland and his principal assistants have had little experience in the building of large masonry or concrete dams previous to the construction of the St. Francis, and apparently did not appreciate the necessity of doing the many things that must be done in order to be certain that the foundations will remain hard, impervious and unyielding.
>
> As a result, various errors were made by an entirely responsible organization confident they were maintaining high standards of accomplishment.

The jury found no indication of negligence on the part of workers who built the dam, concluding that the builders were "deceived and acted in ignorance." In their final stinging reprimand, the jury foreman expressed the panel's evaluation of the catastrophe, a haunting quote that made headlines throughout the nation. "The construction of a municipal dam should never be left to the sole judgment of one man, no matter how eminent."

"The jury recommends there be no criminal prosecution by the District Attorney based on its findings of no evidence of criminal act or intent on the part of the Board of Water Works or any engineer or employee in the construction or operation of the dam. Errors of judgment, absent criminality, caused the tragedy."

In obvious deference by the jury to Mulholland and his supporters, the highly publicized, emotionally charged subject that Asa Keyes had managed so deviously to keep out of the proceedings was addressed only in closing: "Owing to previous attacks on the water system by the use of explosives, consideration has been given to the possibility that this disaster was precipitated by an overt act. While it is undoubtedly *possible* that the destruction of the dam could have been caused by an explosion, no conclusive evidence that such was the case has been brought before us. Even if the failure had been precipitated by this cause, it would not change the situation as far as concerns the existence of the defects that have been described and which were the more probable cause of the disaster."

After the reading of the verdict, a joyous Keyes could not conceal a smile. He saw himself as the prosecutorial David who had felled the Goliath of Los Angeles.

The jury's admission that sabotage was indeed possible, but evidence had not been brought forth to support it, was a hollow victory for Mulholland, now weeping in the hard oak witness chair, his face buried in his hands. To some, doubt had been cast upon the real cause of the collapse forever, but in the eyes of science and the law, the blame had been unequivocally placed on his shoulders.

After their verdict was read, the jurors were dismissed. One by one, they filed out of the room, their faces gray, their eyes downcast.

Bessie Van Norman quietly wept. The room was filled with a bewildered, quiet silence as the crowd slowly dispersed. Outside in the rotunda of the opulent Hall of Justice, the irrepressible Asa Keyes held a long victorious press conference, utilizing the sad occasion to further his political aspirations.

All during the inquest, Mulholland had sat alone taking his own counsel. Now, without speaking, he walked solemnly alone through the crowded courtroom and outside to the street to his waiting car. While the mob of reporters followed him, shouting questions at his back, he got in the black Marmon sedan and rode silently away. For two solitary hours, Mulholland drove aimlessly through the city that had benefited from his genius in its headlong growth from 15,000 souls to almost a million and a half. A genius now discredited and aging.

The Coroner's jury, without the benefit of Pierson Hall's records, or the studies conducted by explosives expert Zatu Cushing, or the findings of the Hercules Powder Company among others, or the affidavits filed by the Stanford professor who expounded the "Stanford Fish" theory, or the affidavits sworn by witnesses who had found the dynamite crate, the map, the tell-tale rope, or who overheard terrorist threats, concluded that dynamite may have caused the dam's failure, but in their own verdict and in discussions with the press afterward, it was abundantly clear that despite the ever-present skepticism concerning the possibility of sabotage, that possibility did not outweigh suspicion that the true cause was the weakness of the rock structure in San Francisquito Canyon.

The office of the District Attorney, Coroner Frank Nance, Mayor George Cryer, members of the city council, board members of the Board of Water and Power Commissioners, leading businessmen and boosters, state officials, federal authorities and the citizens of Los Angeles accepted the jury's verdict without hesitation, and the guilt of William Mulholland was immediately fixed in the minds of the general public.

The verdict proved to be the tragic end to an extraordinary career.

ACCORDING TO PIERSON HALL, politics and geology sadly conspired to suppress the bitter truth surrounding the reasons behind the collapse of the St. Francis. Despite the coroner's verdict, Hall maintained that the dam had been a target of sabotage, and he remained convinced that whoever dynamited the dam perished with it.

Thirty-six years after the tragedy, Hall made his beliefs public. "I am perhaps the only person alive today, who, at that period of time was an official of a policy-making body of the City of Los Angeles," Hall wrote in 1964. In a letter attached to his critical review of author Charles Outland's book on the disaster, Hall revealed his account of what happened behind the scenes, a discussion made possible only by the lapse of three decades.

> I confess to being a part of that "past generation" who did what we thought was, and since proved to be, not only the honorable thing to do, but the only thing to do for the future welfare of this City which has always been and always will be thirsty for water.
>
> Mr. Outland rather seems to accept the skepticism expressed by the *Los Angeles Record* that the dam was dynamited. I think it was. I was at the dam site the day following the break. Shortly thereafter the "self-admitted expert" from Texas examined the face of the blocks lying immediately back of the center standing position. He said that in his opinion dynamite had been used. . . . I called the Hercules Powder Company . . . [they] gave the same opinion. I was still not satisfied with that, so I had a testing laboratory in Los Angeles conduct some experiments, and they came up with the same answer.
>
> All of this was done without fanfare or publicity; indeed nothing was said by me to the City Council concerning these activities. But by the time we came up with the same answer from three unrelated sources, we had long since passed the point where the city of Los Angeles had agreed to assume full responsibility not only for restitution, but for reparations and liability. So that matter was "swept under the rug."

> I do not believe the testing company filed a written report, and nothing but controversy and ill-will could have resulted from any further discussion or disclosures.

Hall was adamant that the city of Los Angeles should immediately accept full responsibility for the tragedy. He concluded that the only proper course for the city was to assume complete and absolute liability for all damages without question.

Hall and other key officials recognized the urgency surrounding passage of the Boulder Dam legislation, and as argument raged in Congress, local advocates, including influential members of the Board of Control, did what they could to bring an end to the publicity and scrutiny surrounding liability for the tragedy. The blame squarely fixed on Mulholland's shoulders added to the quick dissipation of public concern over the catastrophe, and Asa Keyes's zealous onslaught against Mulholland served that end as well.

Hall never informed his fellow city council members about the sensitive research findings concerning the dynamite. He did, however, inform the members of the Board of Water and Power Commissioners, the mayor and key persons at the Department of Water and Power, including William Mulholland and Harvey Van Norman. Hall never made his beliefs public knowledge and never released his findings, telling officials that nothing but ill will would have resulted. Hall and Cushing destroyed their documentation and correspondence involving the dynamite studies to protect what they believed to be the best interests of the city. Pierson Hall desired quick resolution of reparation payments to the victims and did not believe it was advantageous to attempt to "prove the city's innocence."

Mulholland was aware of the evidence which seemingly proved foul play, but nevertheless accepted full responsibility for the tragedy, no doubt conceding at least in his own mind the harm that could reseult from releasing the information to the public. Mulholland had choreographed the delivery of the Boulder Dam and was least likely to jeopardize the project. His acceptance of total blame was in the words of Pierson Hall indicative of Mulholland's "towering courage."

FOLLOWING THE DEPRESSIVE aftermath of the collapse of the St. Francis dam, fearing that economic malaise would settle over the City of Angels and stifle its growth, city fathers and the omnipotent Board of Control searched for a potent cure, a booster shot in the arm for their ailing city. They turned to a promotion that had served them well in the past—a civic parade.

Two weeks after the verdict, the inquest behind them and forgotten, a mind-boggling, five-mile-long parade featuring thirty-two thousand marchers in four divisions of thirty-one high-stepping bands, squads of spit and polish sailors and soldiers, mounted police, Boy Scouts, flying airplanes, pretty girls and children, singing cowboys, dancing Indians, Mexican troubadours, jazz singers, mezzo-sopranos, film stars, and hundreds of decorated automobiles and elaborate floats, one featuring a roaring lion, declawed for the occasion but symbolizing the fierce determined spirit of the city, was turned out for two hundred and fifty thousand gasping Angelenos thronged in the streets to marvel at in wondrous civic pride.

It was, the *Los Angeles Times* proclaimed, "the largest civic procession ever seen west of Chicago," and the largest celebration ever in the history of Los Angeles to date, surpassing by far the flower-strewn, airplane punctuated spectacle in 1924 for the dedication of Mulholland Highway and the earlier dedication of the aqueduct. And it was all there to celebrate the three-day-long dedication of the new Los Angeles City Hall, "a sheer gleaming tower of white symbolizing a new era of progress and accomplishment for the Pacific Southwest" and a warning message to the skyscrapers of New York to beware. Voted in, like the aqueduct, in a bond election, and two years in the making, the building cost the people of Los Angeles $4.8 million, and its towering height of 452 feet soon made the tragic 200-foot center piece of the St. Francis dam a distant memory.

Out of respect, officials had asked William Mulholland to be one of the guests of honor at a dedication luncheon at the Biltmore Hotel, with his faithful friend and ally Mayor George Cryer, and Arthur Eldridge, president of the Board of Public Works. Mulholland sent his regrets and remained at home. That evening, Ruth and Rose Mulholland listened to the live radio broadcast of the spectacle. When President Coolidge pushed a golden telegraph button in Washington, D.C., they rushed to the window to see light beaming into the Los Angeles sky from the huge beacon atop the tower of the new City Hall. The Lindbergh beacon, dedicated to the illustrious aviator, turned silently on its pivot and cast the message of Los Angeles's civic progress and development as an aviation center in a beaming circle 120 miles in diameter. Los Angeles was moving on to new pursuits and new heroes.

MULHOLLAND'S MANY DETRACTORS looked upon the St. Francis disaster as proof of his shortcomings as an engineer, and the city halted construction of his last dam project at San Gabriel after five million dollars had been spent. With amazing speed, Mulholland's beautiful water supply dam in the Hollywood Hills was no longer called Mulholland Dam but quietly and permanently renamed the Hollywood Reservoir.

A new state program for certification of dam construction and safety was abruptly put into force. A new law enacted in 1929 stipulated that all dams must be reviewed by a board of eminent civil engineers and geologists retained by the state engineer before and during construction. This commission subsequently became the Division of Safety of Dams (DSOD) within the State Department of Water Resources, and became one of the first agencies created specifically for dam safety review in the world.

Critics blamed Mulholland's monstrous ego for the disaster, boldly asserting that if he had only retired from his position as Chief of the Department in 1921 at age sixty-six instead of continuing to dominate its every move, none of the tragic events in the San Francisquito Canyon or the Owens Valley would have occurred. "If he had retired this year and left the management of the city's water system to younger hands," William Kahrl observed, "to a mind less inclined to see in every problem an occasion for self-righteous conflict, to someone whose sense of identity was not so completely wrapped up in the city as a whole that he saw each challenge as a personal affront, perhaps the history of Los Angeles's relations with the Owens Valley would have been very different. But Mulholland stayed on, and in only seven years he destroyed his own career, embarrassed the city, devastated the Owens Valley, and undermined the ideal of public water development he had labored so long to establish."

Now, entering a life of self-enforced isolation, Mulholland spent most of his days alone at home except for the occasional company of his sons and daughters. Once limitlessly energetic, outspoken, and vigorous, Mulholland now became listless and withdrawn, although foes maintained that he was without conscience and still the same ruthless old conniver he had always been. Trapped in the brooding despondency that had ensnared his mind, he seemed to forget his many accomplishments.

The world being what it is, a lifetime's effort can be wiped out in a single tragic event, and following the collapse of the St. Francis, the legend obscured the man underneath. Later generations would only vaguely know him from the monuments named for him—Mulholland Highway, Mulholland Dam, San Fernando Valley's Mulholland Junior High School, and later a memorial fountain built at the intersection of Los Feliz Boulevard and Riverside Drive, where Mulholland had lived as a young man working in the crude zanjas.

The characteristic, good-natured combativeness that had given drive and spice to his life seemed to vanish as he entered into his exile, and family members said that he often lacked even the energy to speak to friends and visitors. Catherine Mulholland recalled that following the St. Francis break, detached and aloof, her grandfather appeared like "a silent specter" at gatherings even with his own family, whereas before the disaster his frequent visits to her childhood valley ranch were clamorous affairs and she and the rest of the children would stop playing games and rush inside to greet him by standing at attention and answering his many questions about school. The immense tragedy of the St. Francis weighing so accusatively upon her grandfather's shoulders meant little if anything in the life of a young girl concerned with her Raggedy Ann and games of hop scotch. "In his dark suit, stiff-collared shirt, and cravat, wreathed in the smoke of his ever-present cigar, he was a given in my life, and I loved him as one loves a grandparent, respectfully and unquestionably."

Unlike some of Mulholland's colleagues who declined to visit him during the aftermath of the St. Francis, Raymond Taylor continued to see Mulholland whenever he could during this painful period, both as friend and physician, extending his warm fellowship. Conferring with Mulholland's attending physician, Dr. Anton, Taylor was told that while Mulholland's Parkinson's disease and age-related disabilities could be treated with some degree of success, his severe depression had taken a deep hold of his mind.

During one particular visit, Dr. Taylor found most of Mulholland's teeth and gums in such bad shape that he must immediately visit a dentist if he wanted to save them. When Taylor returned the following day to escort Mulholland to the dentist's office, Mulholland greeted him with a toothless grin, casually relating that after Taylor had departed the day before, he walked into the garage, took a pair of pliers and yanked out all of his decayed teeth one by one without assistance or anesthesia. Needless to say, Taylor was shocked at the incident. For the first time, Taylor wrote, he became keenly aware of the full extent of Mulholland's terrible depression.

But Dr. Taylor was not only concerned with Mulholland's mental condition. Sadly and helplessly, Taylor watched as Mulholland gradually succumbed to the ravages of Parkinson's disease. By 1930, due to the severity of his involuntarily twitching muscles, Mulholland could barely sign his name.

Nine months after the dam's collapse, and only a month before President Coolidge signed the Boulder Canyon Project Act, William Mulholland, now seventy-three years old, after fifty-one years of devoted service, retired from the Department of Water and Power. Scheduled to go to Washington, D.C., in the winter of 1928 to join the President in ceremonies commemorating the signing of the Swing-Johnson bill, the conclusion of his lifetime effort to supply water to his beloved city, Mulholland sent W. B. Mathews and Van Norman in his place.

18

KINGDOM OF ANGELS

There the wicked cease from troubling
and there the weary be at rest.
JOB 3:17

IN AN IRONIC TURN OF EVENTS, as Mulholland wrestled with the demons of his conscience hunkered down in his memento-filled study, District Attorney Asa Keyes was indicted by a Los Angeles grand jury for bribery, conspiracy, and jury tampering in connection with a highly publicized trial involving the Julian Petroleum Corporation.

More than one taxpayer had raised eyebrows at Keyes's expensive, tailored wardrobe, antique-filled Beverly Hills home, spiffy green roadster sports car, gold jewelry, monogrammed golf clubs, unlimited spending cash, and a daughter's steep tuition at Yale, all seemingly paid for with apparent ease on a civil servant's salary.

By early January 1929, Asa Keyes's own deputies were preparing an iron-tight case against their former boss, and by mid-month a high-profile trial against Keyes commenced with state witnesses testifying that thousands of dollars in secret bribes were paid to Keyes in exchange for immunity from prosecution.

After weeks of explosive testimony that garnered national headlines, including accusations that Keyes's moral fiber had been broken by liquor and greed, in February Keyes was sentenced to fourteen years in San Quentin.

The *Los Angeles Times* gleefully reported that "On March 11, shortly after 5 P.M., after twenty-five years as a member of the district attorney's office, escorted by deputy sheriff Frank P. Cochran, Asa Keyes boarded the Southern Pacific Owl for San Quentin. Although he fought to preserve his composure, it was evident that the once powerful public official was a broken man. . . . Keyes wore an old suit and carried only a few personal possessions, including two pipes, a little tobacco, a razor and some stationery."

Keyes's fall from public grace was just as swift as that of William Mulholland, but not as unexpected, nor as personally traumatic. The *Times* failed to run the statement made by Keyes just minutes before his transfer from county jail to San Quentin, and given to the same press corps who, months earlier, had eagerly covered his flamboyant press conferences denouncing William Mulholland.

> In some quarters there is apparently an impression that I am going to San Quentin downcast and broken in spirit. In fact, I have seen statements to this effect in the press lately. Nothing could be farther from the truth. I will enter the State prison with my head held up, looking the world squarely in the face, and will serve my time there, regardless of how long or short it may be, soothed and sustained entirely by the knowledge in my own heart, and the knowledge of my family and friends, that I am absolutely innocent of the crime for which I was convicted. I go to San Quentin firm in the belief that justice will soon come my way. While there my friends can rest assured that I will do the time and not let the time do me.

And true to his word, the time did not "do" him. In October 1931, after only thirty-two months in the "Big House," Asa Keyes, healthy and sound, was paroled and eventually received a full executive pardon from California Governor Rolph. The fact that Keyes during his tenure as district attorney was credited with having sent more than a third of the prisoners in San Quentin to the cells which they were occupying during his stay attested not only to his knack for self-preservation, but also to the extraordinary precautions taken by the warden to protect him.

The ways of fame are more often than not fleeting and capricious. On the occasion of his homecoming, Asa Keyes was greeted by over four hundred people and dozens of reporters on hand at Union Station in downtown Los Angeles, and driven to his Beverly Hills home for an elaborate celebration dinner with family and friends. A week later, Keyes reported to work at his new job as a used car salesman at a Pasadena car dealership. At Union Station a number of fans had tried to slip him their personal checks in hopes of being the very first to purchase an automobile from the once-fearless prosecutor.

ON NOVEMBER 12, 1929, on the occasion of Mulholland's official retirement, Harvey Van Norman was formally installed as chief engineer and general manager of the Department of Water and Power. Van Norman's appointment was applauded by labor groups such as the International Brotherhood of Electrical Workers, and the vast majority of the employees of the department. However, to the many foes of Mulholland's regime desiring a clean slate in the administration of the water department, the appointment was not a satisfactory one.

In an early effort to stave off Van Norman's appointment, confidential memos circulated to Dr. John R. Haynes, President of the Board of Water and Power Commissioners, which offered a disparaging portrait of Van Norman, describing him as lacking integrity, engineering skill, and executive leadership, and branding him a "wily politician who never hesitates to boldly and openly make misstatements." "Unfortunately he has a frank and open manner and is rather captivating to those who do not know his real nature," cautioned one critic, who deplored Van Norman's ties to *Herald, Examiner,* and *Times* editors.

The Los Angeles newspapers and magazines depicted Van Norman as a perfect successor to Mulholland. They described him as an innovative thinker, planner, and doer who comes "ably equipped to step in the large shoes of Chief William Mulholland. . . . Puffing his ever present brier and amicably walking his Boston bulldog on his Sunday evening meditations, he is every inch an able leader, profoundly absorbed in the world around him."

After taking office, one of the first things Van Norman did was retain Mulholland as Chief Consulting Engineer for the Department of Water and Power at a salary of five hundred dollars a month, despite protests from foes charging the appointment was gratuitous, overlooking the fact that Mulholland's long experience in the affairs of the department was an invaluable resource and a bargain for the price. However, as Van Norman was well aware, the accusation was not without foundation. He loathed watching Mulholland waste away listlessly at home, and he hoped the appointment might give him some renewed energy and sense of self-worth.

For a time it worked. Back in the familiar surroundings of the Water Department building at 207 South Broadway and in the warm loyal company of long-time associates, Mullholland sat in on the many day-to-day meetings, his opinions eagerly sought by the young engineers. His dark depression seemed to wane. Then in May 1930, Raymond Rising, the dam engineer who survived the midnight flood by his chaotic ride on a rooftop, filed a series of wrongful death claims against the City of Los Angeles, the Department of Water and Power, and William Mulholland for $175,000 for the deaths of his wife and three young children. Once again, Mulholland was forced to relive the events leading to the catastrophe at St. Francis dam.

Although the city had managed to settle out of court all but nine death claims from the more than four hundred, city attorneys feared that if Rising succeeded in proving the city's negligence, it could open a floodgate of litigation and cost the city millions of dollars. Thus far, the city had managed to keep the genie in the bottle, and what could have cost the city over $20 million in damages, had been quietly and expeditiously settled out of court for roughly $7 million. The Ray Rising trial threatened to change that.

Defense attorneys representing the city immediately attempted to prove the dam collapsed not through negligence by the city or by the chief engineer of the Department of Water and Power, and denied the city was in any way responsible for the deaths of Rising's wife and children. In a complete reversal of the city's position during the coroner's inquest of 1928, city attorney Jess E. Stevens declared that the break in the dam was caused not through faulty engineering, but by an act of God—namely, *an earthquake*. It was far from the city attorney's mind, but Mulholland's previously established guilt so aptly orchestrated by the still-imprisoned Asa Keyes would be indirectly absolved if the act of God theory was proven.

Rising's attorney introduced once again the infamous parlor game, dropping a rock in a glass of water so the new panel of jurors could watch it dissolve, and once again Mulholland wearily denounced its validity.

Rising's attorney then called his principal witness, Professor F. L. Ransome, a geologist at the California Institute of Technology who testified that there was no seismograph record of an earthquake or earth movement of any kind immediately preceding the collapse, and in his learned opinion the collapse was indeed caused by faulty construction on a hazardous site. With nine other similar lawsuits at stake, city attorneys waged a fierce battle to prove the dam collapsed due to an earthquake and presented seven days of testimony by geologists and engineers to corroborate their case.

Following long testimonies about the issue of quakes, Mulholland and the jurors sat through hours of dramatic descriptions of the flood that swept through the Santa Clara Valley on the night of March 12, 1928. The emotional strain on the seventy-five-year-old former chief was plainly acute when he broke down sobbing during Rising's own tearful rehashing of the gruesome details of the events that killed his family.

In his instructions to the jurors, the judge cautioned that if in their opinion the dam collapsed due to an earthquake or earth movement, and that if they found the city had utilized skilled and competent engineering techniques, there could be no liability found against it. He then defined an act of God as a "force of nature that cannot be prevented by human care, skill or foresight, but results from natural causes such as an earthquake, earth movement, lightning, tempests, floods or inundations."

Any hopes William Mulholland had for vindication were soon shattered. After only one hour of deliberation, the jury ruled in favor of Rising and awarded thirty thousand dollars in damages for the deaths of his wife and three children. Again, judgment placed the blame on William Mulholland and, despite the continued attempts of friends and allies like Van Norman to support his rapidly failing ego, he became even more withdrawn.

THE YEAR 1930 found Mulholland troubled and depressed. Fred Eaton's woes ran even deeper although seemingly more readily solvable. Eaton's continued fight to thwart the imminent foreclosure of his coveted Long Valley ranch could be won by simple cold hard cash, but his deteriorating mental condition brought on by the ravages of old age and failing health was accelerated by his obsessive dream of a multimillion-dollar sale of Long Valley to the city of Los Angeles.

In September Eaton offered to sell water rights for fifty-one thousand acre-feet of water to Los Angeles for a staggering $2,300,000. And in behavior described as "irrational and delusional," Eaton boldly declared to the commissioners that his offer was "only good for fifteen days" after which he threatened the water would be sold elsewhere to the highest bidder. In an emotional declaration Eaton stated that his offer represented only 15 cents on the dollar of Long

Valley's actual value, a mere pittance, after all, for the man who conceived and inaugurated the Owens River water-supply project for Los Angeles.

When Van Norman and the commissioners rejected the offer as preposterous, offering the market price of $800,000, Eaton's bombastic reaction, according to witnesses, exhibited signs of paranoia.

Eaton's unreasonableness over Long Valley had generated severe marital troubles. By 1930, the stormy conflicts between Eaton and his wife, Alice, were gleefully highlighted in the pages of Los Angeles newspapers. Convinced her husband had crossed the lines of rational thinking, and in fear of her own economic security, Alice Eaton, long used to a life of ease, filed suit to have him declared incompetent. She asserted that the seventy-five-year-old Eaton was physically and mentally unable to conduct his affairs and had become "insane with an exaggerated idea of the value of the lands he owns in Long Valley." In order to protect the rapidly depleting estate, and Eaton from himself, she sought to be appointed receiver, ousting him from control of the Eaton Land and Cattle Company. Within days, Alice Eaton formally separated from her husband.

An ensuing court battle for the Long Valley property raged between the spouses. Eaton vigorously defended himself against the charges of incompetency and Eaton's attorneys successfully defeated charges that old age and paralytic strokes had incapacitated his judgment.

By January 1932, out of funds, Alice Eaton was forced to dismiss her lawsuit; in March, Fred Eaton was himself broke, and the Long Valley property was submitted into final foreclosure by bank officials.

Alice Eaton's personal holdings were now threatened, and the bank which held the deeds on her rental property and a Los Angeles house acquired as separate property before her marriage, also commenced foreclosure proceedings.

Claiming she was utterly without financial support and forced into desperate personal circumstances, Alice appeared before the Los Angeles City Council begging County Superintendent of Charities William R. Harriman to provide her with money or canned food. Alice Eaton's pathetic plea for charity received front-page coverage by Los Angeles newspapers in the continuing saga of the Owens Valley. Wrote the *Los Angeles Times*:

> . . . So Mrs. Eaton, a native of Los Angeles, once an active club woman accustomed to a life of ease, estranged wife of a former mayor of Los Angeles and a civic pioneer verges on the edge of poverty. "I do not blame him," Mrs. Eaton said of her husband, who lay ill in the Los Angeles home of his eldest son. "He is sick and can't understand my predicament. We do not like to feel that we are separated, although in the eyes of the court we are. But he is unable to do anything.
>
> "The land that we own—what will some day become the Long Valley City Reservoir in Inyo County is worth millions. It is tied up now in receivership. But land like that can't keep me from getting hungry, can't support my sons nor save my home. Today the bank foreclosed on my home—with interest amounting to $200 which I was unable to meet. That was the last straw. I went to see Mr. Harriman although the humiliation was beyond description. And today I had to ask him for help."

The plea of the former mayor's wife generated city-wide expressions of sympathy, but not from the Los Angeles City Council which respectfully declined her request stating the law forbade county aid to applicants with personal property in their names exceeding twenty-five hundred dollars.

Eventually, Alice Eaton lost all of her possessions as well as the property she owned before her marriage. An inventory of the seized items, valued previously by the City Council at over twenty-five hundred dollars, reflects the life of a former mayor's wife and charity club woman whose life somehow had gone terribly wrong: "Two 9 × 12 Wilton rugs, twelve English Chintz curtains, six wrought-iron curtain fixtures, one piano, one Atwater Kent Radio in walnut Console cabinet, three velvet cushions, three Philippine bamboo armchairs, one Philippine Peacock Chair, one Cane desk, two Oak Stickney chairs with Spanish leather cushions, three walnut nest tables, one iron bridge lamp with beaded lapis shade, and various miscellaneous items . . ."

In July Eaton's youngest son, Henry, age twenty-two, published an exclusive article in the *Los Angeles Times Sunday Magazine* about the family's predicament, claiming the earnings from the article would put food on his mother's table for another two weeks.

> The picture of my mother as she is today in comparison with the position she held is very tragic. Here is a worn, sorrowful little woman in a dotted, gingham dress asking the county to help her and her children to exist.
>
> Yet only yesterday, she had been the wife of a prominent mayor of this city: a man who as consulting engineer for the Los Angeles Water Board had received a salary of $100 a day, a cattleman whose grazing lands covered thousands of acres. Money, social position seemed very real things in those days. But where was all that now? Where were the homes too large for our immediate family? Where were the servants, the cars, the chauffeurs, our friends?

The forlorn Henry Eaton may well have concluded that the world had abandoned his unfortunate parents, but a letter written in Mulholland's own firm hand seems to contradict Henry's lament; the unmailed letter was found among Mulholland's papers after his death and seems to indicate that he retained a continuing interest in resolving his former friend's problems. Handwritten on Santa Fe Chief Rail stationery, the letter was dated April 2, 1928, only sixteen days after the collapse of the St. Francis Dam. Despite Mulholland's own bitter heartache following the disaster, he found time to write his ailing friend with advice. Perhaps Mulholland's own troubles reinforced a sense of mortality, and grieving himself, Mulholland felt the need to reach out to his former mentor out of a sense of gratitude or even profound loss. In the letter, Mulholland relays information that he thinks would be helpful to George Khurts, Eaton's agent in negotiations for the sale of Long Valley to the city.

2 April 28

Dear E,

I do not understand yet how I missed seeing George Khurts last Saturday. It was in my mind to say to him something like this. The Board has never evinced the slightest desire to buy the Eaton property—neither the Board nor anyone else in the department has ever recommended it. No price has ever been named or discussed in the department on the subject.

The only price I ever heard of was that suggested by someone of the mayor's office committee—just when, to whom or under what circumstances I do not know. I believe I was away from home at the time and only heard of it afterward. I do not know whether this suggestion was ever reported by the board.

The present is not a proper time to bring the matter up. I do not know what could be done about it now or for some time to come.

Please read this letter to Mr. K personally and ask him to state if he will, whether the figure named by a member of the mayor's committee [who] said $600,000 would be favorably considered. So that if he says it is untimely I may have something definite to act on: I cannot say whether that figure or any other will be paid by the board but I can find out, and thus put an end to the present uncertainty.

Mullholland may have neglected to mail the letter amid the chaos of events or it may reflect thoughts conveyed later.

By year's end, Fred Eaton was finally declared bankrupt. On December 7, 1932, the city council voted its final approval to purchase Long Valley out of receivership for $650,000. Eaton had paid $22.50 an acre for Long Valley in 1905, and the city of Los Angeles, after three decades of violence, obtained it for a paltry $25.00 an acre, forever eradicating Eaton's long dream of riches.

In the eyes of the Board of Control the retribution for Eaton's betrayal of Los Angeles twenty-seven years earlier was now complete. He had committed the unpardonable sin of trying to hold up the predestined, glorious advance of the kingdom of the City of Angels.

AT THE START OF 1934, William Mulholland received an urgent plea from Eaton's family. When the message came, Mulholland put on his hat and hurried out of his house without a word. Eaton lay dying of complications from a final stroke. Ushered to Eaton's bedside at his son Burdick's Los Angeles home, Mulholland quietly greeted his old friend, "Hello, Fred," and took his hand.

The two were then left alone. There is no public record of what the onetime friends spoke of during this, their last visit together, no record of stinging reprimands or gentle words of forgiveness, but the scene was a common one in human affairs—one man on his deathbed and an old requested friend at his side, hands clasped, administering to him. They had suffered great wounding defeats in the twilight of their long careers, one accused of a ruthless pursuit of glory and the other battered by a quest for riches. Both were left in bitter despair.

In those brief minutes behind closed doors, perhaps they restored themselves by healing the long rift between them with fond reminiscences of a happier time.

IN THE FALL OF 1930, Elisabeth Mathieu Spriggs, a student at the University of Southern California, sought a visit with Mulholland for help with her master's thesis on the history of the domestic water supply of Los Angeles. Reluctant to grant any interview, the depressed Mulholland finally did so, and friends recalled that his brief interaction with the sincere and enthusiastic young lady had lifted his spirits, motivating him to heed his appearance, and, for a short time, take a new interest in life. In the course of her interviews, Mulholland had proudly taken her on visits to the many old reservoir and intake sites of the early Los Angeles water system where he worked for Fred Eaton as a ditch tender and later as supervisor.

In her short but glowing biography of Mulholland at the close of her thesis, she listed many of his early achievements in hydraulic engineering and dam building and, of course, recounted the achievement of the Los Angeles aqueduct, quoting Mulholland's remarks at the opening of the Cascades acknowledging Eaton's pivotal role in the establishment of the great man-made river.

There was no mention of the catastrophic St. Francis collapse or of the troubling affairs of Long Valley in Elisabeth Spriggs's thesis, but she poignantly captured in Mulholland's own words the hopeful world of Los Angeles that Eaton and Mulholland shared in pre-aqueduct times.

"The world was my oyster and I was just opening it. . . . Los Angeles was a place after my own heart. It was the most attractive town I had ever seen. The people were hospitable. There was plenty to do and a fair compensation offered for whatever you did. In fact, the country had the same attraction for me that it had for the Indians who originally chose this spot as their place to live. The Los Angeles River was the greatest attraction. It was a beautiful, limpid little stream, with willows on its banks. . . . It was so attractive to me that it at once became something about which my whole scheme of life was woven. I loved it so much."

On March 11, 1934, Fred Eaton, age seventy-eight, died, and that night a brooding Mulholland disclosed to Rose that "For three nights in succession I dreamed of Fred. The two of us were walking along—young and virile like we used to be, yet I knew we were both dead."

Mulholland's hotheaded statement to Van Norman years before that "I'll buy Long Valley three years after Fred Eaton is dead," a vow to deny Eaton a profit in his lifetime, was a prophecy after all. One day before Eaton's death, receivership of the Eaton Land and Cattle Company was terminated and title to Long Valley was transferred to the City of Los Angeles. Once a millionaire, Eaton died leaving an estate estimated at $20,000, consisting of $15,000 in stock from the Eaton Land and Cattle Company and $5,000 in Mono County land. His will divided his estate equally among his wife and six children.

Mulholland's last public speaking appearance was the ceremony for the commencement of construction on the Colorado River Aqueduct at its starting point in Cabazon, near Banning, California. Although still living down the disgrace of the St. Francis dam collapse, he was, by all accounts, the symbolic father of the project, and in attendance were many national public officials and prominent citizens who applauded him profusely when he was asked to speak. Walking with difficulty to the microphone and in his rugged, deep voice he greeted the audience in typically terse Mulholland fashion.

"Well, anything I might say here would be pretty old stuff. I've tramped over these hills since '77 . . . and I'm getting along. I am glad to be of service to you and to this community forever!"

On July 22, 1935, a little over a year after the death of Fred Eaton, William Mulholland died at the age of seventy-nine. His death had been expected for several weeks. Prior to his death, Rose had found her father murmuring in delirium, and asked her brother Perry if he could understand anything he was saying.

After a few minutes, Perry said, "Why, he's giving orders on a ship; he's a sailor again."

At his bedside were sons Thomas and Perry, daughters Rose and Lucille, and his attending physician, Dr. F. L. Anton. Mulholland died quietly in his sleep. Death was attributed to arteriosclerosis, which had caused a severe stroke the past October.

Asa Keyes died three months later at his Rodeo Drive home, three years after his release from San Quentin prison. He was fifty-seven. Harvey Van Norman, possibly the man who knew William Mulholland best, and assuredly Mulholland's most devoted disciple, died in 1954 of a heart attack at seventy-five, only eight weeks after the death of his beloved wife Bessie. They had been married forty-seven years. Van Norman had ultimately built the dam at Long Valley, completed in 1941 with the storage reservoir that sits behind it. Dr. Raymond Taylor, who died in 1958, was the one who lived to see the greatest transformation in the character and beauty of their cherished city.

AS FOR THE BOARD OF CONTROL, whose members profited most from Mulholland's and Eaton's dream, Harry Chandler, who succeeded his father-in-law Harrison Gray Otis as publisher of the *Los Angeles Times*, lived on until 1944. In later years he became known in Los Angeles as "the Archcapitalist." It is rumored that he ordered his personal papers and those of Harrison Gray Otis destroyed a few years before he died. H. J. Whitley invested most of his money in oil in the late 1920s, and purchased forty-eight thousand acres in Central California to begin drilling. The stock market crash of 1929 ruined his fortunes; he suffered a stroke soon thereafter and died June 3, 1931. General Moses Sherman died in 1932. Otto Brant passed away in 1922.

Catherine Mulholland summed up their place in history, writing that Mulholland, Eaton, and the members of the Board of Control represented a blend of spirit, ambition, and idealism, and although found unsympathetic at times, "still one must also acknowledge, however grudgingly, their practical vision, their extraordinary energy and drive, and, finally, their large measure of civic devotion and commitment to their adopted city."

Unlike Fred Eaton, Mulholland on his death left no debts and no unfinished business, and for all the accusations of profiteering from the aqueduct, Mulholland's entire fortune, as revealed in his last will and testament, derived from his earnings as a salaried employee for the city of Los Angeles, lucrative free-lance consulting work, and prudent investments in real estate—not from land speculation linked with the members of the Board of Control.

His estate, valued at $700,000, consisted of the 640-acre Mulholland Ranch in what is now called Chatsworth, forty-eight hundred shares of capital stock of the Mulholland Orchard Company, various oil leases, stocks, bonds, and real estate in Montebello and Los Angeles, including the St. Andrew's Place home, which he left to his eldest daughter, Rose.

Mulholland's estate, held carefully controlled in an elaborate trust, provided handsomely for each of his five children. When he died, inside the pocket of his trousers was five thousand dollars in cash tied in a rubber band. His most cherished possession was his inscribed gold retirement watch, worth thirty-five dollars, a gift of the Los Angeles Department of Water and Power.

When news of Mulholland's death was received, the army of engineers and workmen laboring on the construction of the Colorado River Aqueduct ceased work for ten minutes in silent tribute. The flow of water in the Owens River aqueduct was stopped briefly as it came from the intake in Owens Valley. Then the great clamor of earth, machines, and men took up again, inexorably advancing the hard-won water toward the thirsty city of Los Angeles.

EPILOGUE

VERDICT RE-EXAMINED

THE VERDICT AGAINST William Mulholland in the collapse of the St. Francis Dam remained unchallenged for sixty-four years. In October 1992, a new examination of the dam's failure concluded that, given the geological knowledge of the time, Mulholland was innocent of professional negligence in the dam's construction. Findings published by the Association of Engineering Geologists reveal that Mulholland went to his grave shouldering far too much blame for the catastrophe.

After studying the dam for nearly fifteen years, J. David Rogers, a distinguished consulting engineer who investigates dam failures, concluded that the dam collapsed because its eastern edge was anchored into an ancient (paleolithic) landslide impossible to detect. On the night of March 12, 1928, the slide partially reactivated, plowing into the structure like a bulldozer blade, causing a rapid chain-reaction.

Rogers gained a critical piece of the puzzle when he uncovered the testimony of the two mystery eyewitnesses who remembered driving on the dirt canyon road near the dam less than one hour before the collapse and noticed that the road had dropped "at least twelve inches, just upstream of the dam's east abutment."

"When I read that, I came out of my seat," Rogers said. "That tells you the abutment was beginning to drop and thrust against the back of the dam, causing the reservoir to tilt." The ominous sounds of a landslide heard by motorcyclist Ace Hopwell, who had stopped on the road that night, and the "extra-normal sounds of high water discharge" that awakened Ray Rising minutes before the dam's collapse also substantiate Rogers's eastern abutment landslide theory. If Rogers's newest assessment is correct, the dam fell because of an ancient landslide, a condition predicated on geological principles not yet formulated. The tragedy was caused both by an undefined geological condition and an act of God.

Had Mulholland's exoneration come during his lifetime, he would have moved into semi-retirement, world-famous and content, adopting the role of senior statesman whose opinions would have been widely sought. But the phone calls, letters, requests for speeches, invitations, and consulting fees stopped suddenly and irrevocably in March 1928, and a satisfying retirement following a life devoted to public service was denied the chief engineer.

Mulholland was a giant of American engineering who played key roles in the construction of three of America's seven wonders of engineering as defined by the American Society of Civil Engineers: the Colorado River Aqueduct, the Panama Canal, and Hoover Dam. Mulholland's water projects continue to function and serve their purpose years later, probably the greatest testimony to his abilities.

Some argue that Mulholland was guilty of believing in his own infallibility, corrupted by a false sense of power and authority. Some have argued that city officials entrusted him with so much power due to his earlier successes that there were no critical evaluations of his plans. "By that time," commented one member of Mulholland's family, "he felt he had a special dispensation that he'd always be right . . ."

Whatever the truth behind the failure of the dam, one fact is abundantly clear. He was a man of legendary accomplishments, and his downfall, when it came, was of equal magnitude.

IRONIC END TO THE WATER WARS

SINCE THE LATE 1930S, the water wars between the city of Los Angeles and the people of the Owens Valley have been the backdrop for novelists, screenwriters, and filmmakers. In fictionalized form the story is already well-known to millions as the basis for the 1974 Academy Award–winning film *Chinatown,* in which character Hollis Mulwray—whose name is a play on Mulholland—is murdered for his honesty in a complicated plot involving water dumping, incest, and land speculation. The brilliant film is probably responsible for misinforming the public about the chronology of events and promoting the lingering widespread belief in an aqueduct conspiracy.

Historian John Walton observed that *Chinatown* may have contributed to the refueled protest against the city of Los Angeles that broke out in the Owens Valley in the 1970s. Violence returned to the Alabama Gate on September 16, 1976, when an explosion near Lone Pine ripped apart a spillgate of the aqueduct, forcing a shutdown of the system for two full days. The explosion was allegedly in response to a heated dispute between Owens Valley residents and the Department of Water and Power over diversion of water through underground pumping. Mirroring the events decades earlier, the dynamite blasts again made the front pages of Los Angeles papers. In a macabre form of protest in 1978 an arrow tied with sticks of dynamite pierced the breast of a commemorative statute of William Mulholland in a Los Angeles park.

Twenty years after the completion of Mulholland's aqueduct, the farming region of the Owens Valley had turned into a desolate, high-desert environment. Sagebrush replaced once-vibrant wetlands, and alkali dust drifted across the dry bed of Owens Lake. The Owens River that thundered southward during Mulholland's buckboard trip with Fred Eaton in 1904 was reduced in places to a small stream. Although the population of the Owens Valley dwindled, a few stubborn ranchers held onto their land, and with equal stubbornness continued their legal search for the justice they felt was due them. Their lawsuits and legal tactics were thwarted at every turn by the considerable political and legal resources of Los Angeles and the Department of Water and Power.

Then, finally—and unexpectedly—in the 1980s, the delicate Mono Lake ecosystem, a wildlife sanctuary in the Owens Valley, began to disintegrate, prompting a far-ranging debate. As a result of national attention coming in an era of ecological concern, a new agreement was reached. As Mulholland had feared, the valley residents managed to close off a large portion of the flow of the Owens River to the aqueduct to Los Angeles.

On October 16, 1991, after years of haggling and bitter litigation, the Los Angeles Department of Water and Power signed an agreement with the Inyo County Board of Supervisors that limits, once and for all, the amount of water that can be pumped to Los Angeles. The historic agreement requires Los Angeles to pay Inyo County two million dollars a year to help mitigate earlier environmental damage and offset low tax assessments on local land owned by the Department of Water and Power. The agreement also requires Los Angeles to spend ten million more to re-establish a trout fishery in nearly fifty miles of the lower Owens River. In exchange, Inyo County has dropped its challenges to Los Angeles's water rights.

The significance of *Chinatown,* John Walton wrote, is that despite factual inaccuracies, it captured the "deeper truth of the rebellion"—the belief that metropolitan interests illegally and immorally appropriated the Owens Valley for their own expansionary purposes. The film's incest subplot was a metaphor for the perceived "rape of the Owens Valley," and sexual symbolism for the "vile association of money and political power."

The truth, garnered from examination of Department of Water and Power records and Mulholland's personal papers, reveals no grand conspiracy or contrived drought. Los Angeles voters overwhelmingly approved the aqueduct bonds ten to one. The Board of Control's land syndicate did purchase real estate in the San Fernando Valley and later profited handsomely, but their acquisitions were common knowledge, and apparently ignored by the public who shared in the booster spirit of Los Angeles in those early years. J. B.

Lippincott's clear conflict of interest between the city and valley interests as an official of the Federal Reclamation Service and Fred Eaton's unsuccessful speculation at Long Valley were noted and questioned at the time.

The crime Mulholland may have been guilty of was unyielding stubbornness in response to great provocation. By staunchly refusing to give in to Fred Eaton's extortion and by refusing to acknowledge the rightful fury of the Owens Valley people, he effectively set the stage for his own undoing—the tragic collapse of the St. Francis Dam in 1928. If a settlement, even a profitable one for Eaton, had been hammered out in the early years and the reservoir constructed at Long Valley, not only would the Owens Valley have prospered, but the urgent need for a southern storage reservoir built on the unstable rock of the Santa Clara Valley would have been eliminated. Mulholland's intractable position was understandable, but ultimately disastrous.

A dam at Eaton's venerable Long Valley was eventually constructed, supervised by Harvey Van Norman in 1941, after both William Mulholland and Fred Eaton were dead. The reservoir, impounding 163,000 acre feet of water, was named Lake Crowley in honor of the desert padre, Father John J. Crowley, who devoted much of his time to healing the division between the northern desert towns and the city to the south.

MULHOLLAND'S PROPHECY

WHAT IS THE LEGACY of William Mulholland? Critics argue that Los Angeles would have been better off had the population and prosperity envisioned by Mulholland never been promoted. There is some support for this view. Today, Mulholland's tree-lined home at St. Andrew's Place is in the center of a run-down urban expanse, a stone's throw from the site of looting and arson in Korea Town during the 1992 civil disturbance in Los Angeles. The Mulholland Orchard in the San Fernando Valley no longer exists; in its place is a Kmart discount department store. Because of smog, the magnificent vistas from the twisting Mulholland Highway can be seen only on a few clear days each year.

But there is another side to the legacy. The cup is overflowing in America's second city as never before. Tides of immigrants from other parts of the world have forever altered its social landscape. Foreign investment is soaring, and Los Angeles County, previously not considered an industrial stronghold, now leads the nation in manufacturing exports. The city is a leader in architectural innovation, education, and the arts. The phrase "melting pot" has now been revised by city leaders who speak of a "giant salad bowl," of trade, capital, labor, and cultures. Futurists place Los Angeles at the epicenter of the Pacific Rim—a key zone for international trade in the next century.

Mulholland's ingenuous masterpiece, the Los Angeles Aqueduct, did indeed make possible the waves of immigration that brought the burdens and costs of a city growing too fast for its planners, but the resultant diverse population will prove an advantage in the global village of the future.

Without Mulholland's aqueduct, Los Angeles would have been limited in growth, to probably not more than 250,000 people. The aqueduct made possible a metropolis where natural conditions forbid it. The annexation of the San Fernando Valley, a direct result of the aqueduct, instantly made Los Angeles the largest city in the world in terms of geographic size, and from that moment forward, as Mulholland predicted in his celebratory toast in 1913, the citizens of Los Angeles were to be a people "doomed to success."

Mulholland's malapropism was prophetic—eight decades later, the sprawling giant has grown to a region over 100 miles in diameter sustaining a population of 14.5 million and the city of Los Angeles is expected to match the population of America's largest city, New York, near the year 2010. But the promised success now decays under the burden of urban sprawl, violence, unbreathable air, and rationed water.

The Los Angeles populace enters 1993 facing a seventh year of drought and a further decline in their quality of life. Even in wet years the region is dry, receiving on average one-third the amount of rain as New York and half that of Chicago.

In May 1990, Los Angeles instituted mandatory water rationing, requiring households to cut water use by ten percent. In spite of rationing, and even with the possibility of a new aqueduct, it is unlikely the region's critical water problems can be solved. As the twenty-first century approaches, the same problems that challenged William Mulholland confront the city again.

Present sources of water are not adequate and will be further reduced as Arizona and Nevada exercise their legal rights to Colorado River water in coming years. In addition, the problem of drought years has never been adequately addressed. Los Angeles faces long- and short-term water problems of confounding complexity.

The city of Los Angeles, one researcher commented, "would do well to relinquish its tropical self-image and come to terms with its arid climate." Enforced conservation is likely to become the only solution. Even when the drought ends, Los Angeles will have to endure restrictions on lawn-watering, and car-washing, and will have to forego the luxury of abundant swimming pools, lush landscaping and other water extravagances. Conservation devices in addition to the low-flow faucets, toilets, and shower heads already required will be mandatory. Water bills will increase, and penalties for overuse will stiffen. A massive education program is needed to teach consumers to adapt themselves to a desert environment.

The populace, baffled by the complexity of water issues and misled by public leaders enamored of ever-expanding development, is prone to insist no water shortage exists. The history and fable of Los Angeles's water campaigns has left the public suspicious, uncertain of the truth and leery of scare tactics. What is needed is another visionary Mulholland to genially deliver the bad news and rally support for a creative solution, for there are no more rivers to bring to the desert.

NOTES

CHAPTER 1. GENESIS

11 "It was as if Boston . . . twentieth century.: Kevin Starr, *Material Dreams: Southern California Through the 1920s* (New York: Oxford University Press, 1990), p. 56.

CHAPTER 2. HAND OF BETRAYAL

16 Lippincott shrewdly told . . . water wealth: Los Angeles Aqueduct Investigation Board, *Report,* 31 August 1912, pp. 42, 53. See also *Los Angeles Examiner*, 25 August 1905, p. 1.

17 Were it not . . . my hands.: *Los Angeles Examiner*, 17 August 1905, p. 1.

17–18 To celebrate, Eaton . . . future enterprise.: *Pasadena Star News*, 5 August 1905, p. 1.

18 "I have not received . . . it for.": *Los Angeles Examiner,* 5 August 1905, p. 1.

18 "Why, yesterday . . . half a million.": *Ibid.*

20 In June 1906 . . . Alice B. Slosson.: *Los Angeles Herald*, 24 June 1906, p. 3.

20 On hearing news of the marriage . . . congratulations: William Mulholland to Fred Eaton, 18 July 1906, WP04–21:19. William Mulholland's Office Files. General Manager's Office Historical Records. Department of Water and Power, Los Angeles.

CHAPTER 3. SWEET STOLEN WATER

22 "The last spike . . . secured.": *Los Angeles Times*, 29 July 1905, p. 1.

22 "Titanic project . . . River!": *Ibid.*

26 "If we could only . . . own meaty fist.: *Los Angeles Examiner*, 5 September 1905, p. 14.

26 "If Los Angeles . . . water supply!": John Russell McCarthy, "Water: The Story of Bill Mulholland," *Los Angeles Saturday Night*, 11 December 1937, p. 31.

28 Thanks to . . . had been secured.: *Los Angeles Examiner*, 24 August and 28 August 1905, p. 1.

30 "if you leave . . . care of itself.": *Ibid.*, p. 28.

30–31 In 1902 . . . carried in Mulholland's head.: Joseph B. Lippincott, "William Mulholland: Engineer, Pioneer, Raconteur," *Civil Engineering* 2, no. 2 (1941): 161.

32 "Owens River . . . contented people.": *Los Angeles Times*, 8 September 1905, p. 1.

CHAPTER 4. FAITHFUL SERVANT

CHAPTER 5. NOISE OF MANY WATERS

CHAPTER 6. BLOOD OF SACRIFICE

For these three chapters, details concerning Mulholland's construction of the Los Angeles Aqueduct can be found in the following:

Aqueduct Clipping File, Doheny Library, University of Southern California, Los Angeles.

Aqueduct Investigation Board, *Report of the Aqueduct Investigation Board to the City Council of Los Angeles*, 31 August 1912.

Janet Beneda, "The Los Angeles Aqueduct: The Men Who Constructed It," Thesis, institution unidentified, May 1974. Currently in Los Angeles Department of Water and Power Library.

Olga Brennecke, "How Los Angeles Built the Greatest Aqueduct in the World: A Story of Interesting Municipal Activity," *Craftsman*, November 1912, pp. 188–96.

California Railroad Commission, *Los Angeles Aqueduct: General Construction and Auxiliary Costs*, compiled by O. E. Clemens, 1 February 1915.

Frederick C. Cross, "My Days on the Jawbone," *Westways*, May 1968, pp. 3–8.

Department of Public Service, *Complete Report on Construction of the Los Angeles Aqueduct* (Los Angeles, 1916).

Early Water System Records. General Manager's Office Historical Records. Department of Water and Power, Los Angeles.

William Grimes, "Chunnel Vision," *New York Times Magazine*, 16 September 1990.

Burt A. Heinley, "The Longest Aqueduct in the World," *Outlook*, 25 September 1909, pp. 215–20.

______, "Water for Millions," *Sunset* 23, no. 12 (1909): 631–38.

______, "Carrying Water Through a Desert," *National Geographic*, July 1910, pp. 568–96.

______, "Aladdin of the Aqueduct," *Sunset* 26, no. 4 (1912): 465–67.

______, "An Aqueduct Two Hundred and Forty Miles Long," *Scientific American*, 25 May 1912, p. 476.

______, "Restoring the Los Angeles Siphon," *Municipal Journal*, 7 May 1914, pp. 633–35.

______, "Aqueduct Outlet Cascades," *Engineering News*, 2 September 1915, pp. 455–56.

W. W. Hurlburt, "William Mulholland, Man and Engineer," *Western Pipe and Steel News* 2, no. 1 (1925): 5–12.

William K. Jones, "The History of the Los Angeles Aqueduct," thesis, University of Oklahoma, 1967.

"Los Angeles Aqueduct," *Building and Engineering News*, 25 August 1915, p. 6.

William Mulholland's Office Files. General Manager's Office Historical Records. Department of Water and Power, Los Angeles.

Henry Z. Osborne, "The Completion of the Los Angeles Aqueduct," *Scientific American*, 8 November 1913, pp. 364–65.

Gertrude Pentland, "Los Angeles Aqueduct with Special Reference to the Labor Problem," thesis prepared as "Report to Department of Economics," institution unidentified, May 1916. Currently in Los Angeles Department of Water and Power Library.

Richard Prosser, "William Mulholland, Maker of Los Angeles," *Western Construction News* 1, no. 8 (1926): 43–44.

W. W. Robinson, "Myth Making in the Los Angeles Area," *Southern California Quarterly*, March 1963, p. 82.

Roscoe E. Shrader, "A Ditch in the Desert," *Scribner's*, May 1912, pp. 538–50.

William R. Stewart, "A Desert City's Far Reach for Water," *World's Work*, November 1907, pp. 9538–40.

"The Great Aqueduct and What It Means to Los Angeles," *Los Angeles Financier*, 7 October 1910, p. 3.

"Nine Miles of Siphons," *Literary Digest* 46, no. 9 (1913): 452.

Los Angeles Times, 12 July, 21 July, and 30 August 1908; 26 June and 12 September 1909; 27 February and 28 February 1911; and 18 June, 19 June, and 28 August 1912.

Los Angeles Herald, 26 February and 27 February 1911.

Los Angeles Examiner, 18 June and 19 June 1912; 19 January 1913.

Los Angeles Daily Times, 13–19 June 1912; 1 January 1913.

Graphic, 25 July 1908; 10 April 1909.

CHAPTER 7. DELIVERANCE

88–89 This is . . . Angels.': Allen Kelly, *Pictorial History of the Aqueduct* (Los Angeles: 1913), p. 1.

97 "Nothing . . . shoulders": Catherine Mulholland, *The Owensmouth Baby: The Making of a San Fernando Valley Town* (Northridge, Calif.: Santa Susana Press, 1987), p. 72.

97 "Your . . . forever": *Los Angeles Times*, 6 November 1913, p. 1.

98–99 "The magnitude . . . good": *Ibid.*

101 Board of Control: C. Mulholland, *The Owensmouth Baby*, p. 9.

102 "There are certain combinations . . . Chandler.": William G. Bonelli, *Billion Dollar Blackjack: The Story of Corruption and the Los Angeles Times* (Beverly Hills, Calif.: Civic Research Press, 1954), p. 22.

103 Huntington despised . . . before.: William L. Kahrl, *Water and Power: The Conflict Over Los Angeles' Water Supply in the Owens Valley* (Berkeley: University of California Press, 1982), p. 99.

CHAPTER 8. PRODIGAL DAUGHTER

105 On April 28 . . . cervix.: Certificate of Death, County of Los Angeles.

105 "devoted loyal wife . . . Mulholland.": *Los Angeles Times*, 30 April 1915, p. 3.

107 Author . . . city.: Joseph Campbell, *The Power of Myth* (New York: Doubleday, 1988), p. 136.

109 "Well goddamn it . . . them.": Robert William Matson, *William Mulholland: A Forgotten Forefather* (Stockton, Calif.: Pacific Center for Western Studies, 1976), p. 32.

111 On June 15 . . . fire.: *Los Angeles Examiner*, 17 June 1915, p. 1.

113 According to . . . papers.: *Los Angeles Times*, 21 September 1918, p. 1.

113–14 The handsome . . . 1950.: Last Will and Testament of Clara S. Sloan, 27 January 1947, Los Angeles County.

114 One year . . . Southwest.": *Pasadena Evening Post*, 30 October 1919, p. 1.

114 The day after . . . Lillian.: Petition for Appointment of Guardian by William Mulholland, In the Matter of the Estate and Guardianship of Lillian E. Sloan, 31 October 1919.

115 Benjamin . . . politics.: John Steven McGroarty, *History of Los Angeles County* (Chicago: American Historical Society, 1923), p. 61.

116 "trick and device": *Los Angeles Times*, 27 March 1920, p. 1.

116 Clara Sloan . . . kidnapping.: *Los Angeles Examiner*, 25 March 1920, p. 1.

116 Lucille avoided arrest . . . consent.: Petition for Revocation of Letters of Guardianship by Lucille M. Strang, In the Matter of the Estate and Guardianship of Lillian E. Sloan, 14 February 1920.

116 Four days . . . grandfather.: *Los Angeles Times*, 1 April 1920, p. 1.

116–17 This time . . . 1921.: *Los Angeles Times*, 26 February 1921, p. 3.

117 Amazingly, less . . . Lillian.: *Los Angeles Times*, 27 June 1921, p. 1.

119 To buy . . . Strathearn.: Catherine Mulholland, *The Owensmouth Baby: The Making of a San Fernando Valley Town* (Northridge, Calif.: Santa Susana Press, 1987), p. 75.

120 The groom's father . . . years.: *Ibid.*, 170–71.

121 When Perry . . . future.: *Los Angeles Weekly*, 20 July 1990, p. 39.

CHAPTER 9. EDEN

125–26 They gave him . . . basket.: Walter V. Woehlke, "The Rejuvenation of San Fernando," *Sunset* 32, no. 2 (1914): 357.

127 The selling . . . sale.: Kevin Starr, *Material Dreams: Southern California Through the 1920s* (New York: Oxford University Press, 1990), p. 71.

128 Harvey Van Norman . . . secretary.: Catherine Mulholland, *The Owensmouth Baby: The Making of a San Fernando Valley Town* (Northridge, Calif.: Santa Susana Press, 1987), p. 11.

128–29 Despite . . . 1930s standards.: Estate of William Mulholland, Probate File no. 151991, Los Angeles County.

131 Chafee had turned . . . enough!": Raymond Taylor, *Men, Medicine, and Water: The Building of the Los Angeles Aqueduct 1908–1913* (Los Angeles: Friends of LACMA Library, 1982), p. 157.

CHAPTER 10. WARS IN HEAVEN

135 Eaton had no . . . Long Valley.: Inquiry: What Rights Does Los Angeles Have in the Eaton Lands in Long Valley? Undated memorandum to William Mulholland, WP06-1:8, Owens Valley Historical Records. Real Estate Division Historical Records. Department of Water and Power, Los Angeles.

135 "I'll buy the Long Valley . . . dead.": "The Owens Valley Revolt," *The Story of Inyo* (Bishop: Chalfant Press, 1922), pp. 382–83.

136 Verbal attacks . . . continued.: Petition, 1905, in the National Archives. Department of the Interior, Reclamation Service. Record Group 527, Owens Valley Project.

136 Mary Austin . . . right.: Mary Austin, "The Owens River Water Project," *San Francisco Chronicle*, 5 September 1905, p. 7.

136–38 "President . . . Parker": Lesta V. Parker to Theodore Roosevelt, 15 August 1905, in the National Archives. Department of the Interior, Reclamation Service. Record Group 527, File no. 115, Owens Valley Project.

138 "Booming Los Angeles . . . Era.: John Walton, *Western Times and Water Wars: State Culture and Rebellion in California* (Berkeley: University of California Press, 1992), p. 151.

138–39 "Ten years ago . . . desolation.": Morrow Mayo, *Los Angeles* (New York: Knopf, 1933), pp. 245–46.

140 Private eye . . . convict them.: *Los Angeles Evening Express*, 24 May 1924, p. 1.

140–41 In one . . . say.: *Los Angeles Examiner*, 28 May 1924, p. 1.

143 "There are no more . . . Colorado.": Los Angeles Board of Public Service Commissioners, *Twenty-second Annual Report for the Fiscal Year Ending June 30, 1923*, p. 68.

143–44 "You're from the Park . . . to destroy.": Mark Reisner, *Cadillac Desert: The American West and Its Disappearing Water* (New York: Viking, 1986), p. 95.

145 One famous press . . . mystique.": Box 123, John Randolph Haynes Papers. Department of Special Collections, University of California, Los Angeles.

145–46 "The preliminary . . . turkey cocks": *Los Angeles Examiner*, 23 July 1925, p. 5.

CHAPTER 11. UNBOWED

147 In July . . . head.: Certificate of Death, Los Angeles County; See also *Los Angeles Times*, 16 July 1924, p. 21.

149 "If Bill . . . was so.": Burt Heinley, "Aladdin of the Aqueduct," *Sunset* 26, no. 4 (1912): 465.

153 An orchestra . . . will undo.: *Los Angeles Daily News*, 20 November 1924, p. 1.

CHAPTER 12. RETRIBUTION

155–57 The troubles in . . . at Owens Valley.: *Los Angeles Times*, 24 December 1922, p. 1, and 28 December 1924, Part 2, p. 1; *Los Angeles Times Magazine*, 1 January 1924, pp. 3–5; *Los Angeles Times*, 13 April 1924, p. 1.

158 "incident at . . . world,'": Chalfant to Austin, 8 November 1932, Mary Austin Papers. Henry E. Huntington Library, Special Collections, San Marino, Calif.

165 According to . . . Watterson enterprises.: William Kahrl, "The Politics of California Water," *California Historical Quarterly* 55, no. 1 (Spring 1976): 181.

168 Burdick's wife . . . concerned.: Mulholland to W. B. Mathews, 13 February 1928. WP06–1:8, Owens Valley Historical Records. Real Estate Division Historical Records. Department of Water and Power, Los Angeles.

169 "Going to see . . . without his interference.": H. A. Van Norman to W. B. Mathews, 13 February 1928. WP06–1:8, Owens Valley Historical Records. Real Estate Division Historical Records. Department of Water and Power, Los Angeles.

CHAPTER 13. RIVERS OF HADES

177 "Johnny . . . to live?": *Los Angeles Times*, 15 March 1928, p. 3.

179–80 "Dear Sir . . . find.": Mrs. Ann Holsclaw to William Mulholland. WP04–10:17, Early Water System Records, Department of Water and Power, Los Angeles.

For details of the collapse of the St. Francis Dam, see:

Los Angeles Herald, 13–19 March 1928

Los Angeles Evening Express, 13–19 March 1928

Los Angeles Daily News, 13–19 March 1928

Los Angeles Examiner, 13–19 March 1928

Los Angeles Times, 14–22 March 1928

Los Angeles Illustrated News, 14 March 1928

New York Times, 14–21 March 1928

Charles F. Outland, *Man-made Disaster: The Story of the Saint Francis Dam* (Glendale, Calif.: Arthur H. Clarke Co., 1977).

WP04–10:12 through WP04–10:18, Early Water System Records, Department of Water and Power, Los Angeles.

Tim St. George, "That Dreadful Night," *Westways,* March 1961, p. 14.

Ted Thackeray, Jr., "L.A.'s Worst Disaster," *Los Angeles Herald-Examiner,* 10 March 1963, p. B-1.

CHAPTER 14. BREATH OF VENGEANCE

182 Twenty-four hours . . . KILL MULHOLLAND!: Charles F. Outland, *Man-made Disaster: The Story of the Saint Francis Dam* (Glendale, Calif.: Arthur H. Clarke Co., 1977), p. 167.

183 "Mulholland's Heart . . . Mulholland.: *Los Angeles Times,* 14 March 1928, p. 2.

184 In a benefit . . . Ingenues.: *Los Angeles Herald,* 19 March 1928, p. 1.

187–97 "Did you Dr. Webb . . . laughter.: Los Angeles County Coroner, *Transcript of Testimony and Verdict of the Coroner's Jury in the Inquest over Victims of the Saint Francis Dam Disaster,* April 1928, Book 26902.

For details of the inquest, see:

Los Angeles County Coroner. *Transcript of Testimony and Verdict of the Coroner's Jury in the Inquest over Victims of the Saint Francis Dam Disaster,* Book 26902, April 1928.

Los Angeles Examiner, 14–21 March 1928

Los Angeles Herald, 14–21 March 1928

Los Angeles Record, 13 April 1928

Los Angeles Times, 14–21 March 1928

Literary Digest, 14 April and 21 April 1928

"Catastrophe in California," *Time,* 26 March 1928, p. 13

"The St. Francis Dam," *Science,* 23 March and 13 July 1928, p. vii

"California's Dam—Soft Rock," *Outlook,* 11 April 1928, p. 563

CHAPTER 15. PERSECUTION

202–3 Respected city councilman . . . this conclusion.: Pierson M. Hall, "Review of the First Edition of *Man-made Disaster*," *California Historical Society Quarterly* 45, no. 1. (1964): 288.

204 "Stanford Fish": Charles F. Outland, *Man-made Disaster: The Story of the Saint Francis Dam* (Glendale, Calif.: Arthur H. Clarke Co., 1977), p. 217.

204 While Mulholland . . . at this time.": Pierson Hall Papers. Ventura County Museum of History and Art, Ventura, Calif.

204–5 So passionate was . . . accused.: Box 126, John Randolph Haynes Papers. Department of Special Collections, University of California, Los Angeles.

214 To Keyes's . . . 25 percent.: See *Hollywood Dam News,* 4 August 1928; *Los Angeles Times,* 25 April 1928; and *Los Angeles Herald,* 25 July 1928.

214 The Chief's keen. . . as sound.: Joseph B. Lippincott, "William Mulholland: Engineer, Pioneer, Raconteur," *Civil Engineering* 2 (March 1941): 105.

CHAPTER 16. JUDGMENT

219 "What's the . . . gone.": Robert William Matson, *William Mulholland: A Forgotten Forefather* (Stockton, Calif.: Pacific Center for Western Studies, 1976), p. 64.

220 He told Taylor . . . never-ending list.: Unpublished Memoirs of Raymond G. Taylor, ("Recollections of 60 Years of Medicine in Southern California"), Los Angeles County Medical Association Library, Los Angeles.

231–32 Sedgwick declared . . . verdict remained.: Report to Asa Keyes, *The Failure of the Saint Francis Dam,* by Allan E. Sedgwick, Louis Z. Johnson, Walter G. Clark, and Charles T. Leeds, March 1928.

232 Toastmaster . . . brief for the people.": C. A. Dykstra, "A Tribute to William Mulholland," *Intake* 5, no. 12 (1928): 2–4.

CHAPTER 17. REDEMPTION

235 "a buzzard of . . . interests.": "Saint Francis Dam Disaster Invites Evil Politics," *Graphic*, 17 March 1928, p. 16.

236 "The people feel . . . the gnat.": *Ibid.*

237–38 Responsibility . . . tragedy.": Verdict of the Coroner's Jury in the Saint Francis Dam Disaster, April 1928.

240–41 I confess . . . disclosures.: Hall to Servin, 19 October 1964, Charles Outland Papers. Ventura County Museum of History and Art, Ventura, Calif.

242 "largest civic procession . . . Chicago,": *Los Angeles Times*, 26–29 April 1928, *Los Angeles Herald*, 26–29 April 1928, and *Los Angeles Examiner*, 26–29 April 1928.

243–44 "If he had . . . so long to establish.": William L. Kahrl, *Water and Power: The Conflict Over Los Angeles' Water Supply in the Owens Valley* (Berkeley: University of California Press, 1982), p. 232.

244 "a silent specter.": Catherine Mulholland, "Recollections of a Valley Past," in Gary Soto, ed., *California Childhood: Recollections and Stories of the Golden State* (Berkeley: Creative Arts Book Co., 1988), p. 181.

245 Unlike some of . . . terrible depression.: Raymond G. Taylor, M.D. Unpublished Memoirs, ("Recollections of 60 Years of Medicine in Southern California"), Los Angeles County Medical Association Library, Los Angeles.

CHAPTER 18. KINGDOM OF ANGELS

249–50 In an early . . . one critic: Box 128, John Randolph Haynes Papers. Department of Special Collections, University of California, Los Angeles.

250 Then in May . . . young children.: *Ray E. Rising* v. *City of Los Angeles*, Case no. 273625-628. See also *Los Angeles Times*, 26 May, 27 May, 29 May, and 6 June 1930.

255 The picture of . . . friends?: *Los Angeles Times Sunday Magazine*, 24 July 1932, p. 4.

256 Dear E . . . uncertainty.: William Mulholland to Fred Eaton, 2 April 1928. WP06–1:8, Owens Valley Historical Records. Real Estate Division Historical Records. Department of Water and Power, Los Angeles.

258 "The world was my oyster . . . so much.": Elisabeth Mathieu Spriggs, "The History of the Domestic Water Supply of Los Angeles." Master's thesis, University of Southern California, Los Angeles, 1931, p. 67.

260 "still, one must . . . adopted city.": Catherine Mulholland, *The Owensmouth Baby: The Making of a San Fernando Valley Town* (Northridge, Calif.: Santa Susana Press, 1987), p. 131.

261 Mulholland's estate . . . Department of Water and Power.: *The Matter of the Estate of William Mulholland,* File no. 151991, Los Angeles County.

EPILOGUE

263–64 a critical piece . . . landslide theory.: J. David Rogers, "Reassessment of the St. Francis Dam Failure," *Engineering Geology Practice in Southern California* (Belmont, Calif.: Star Publishing, 1992), pp. 639–666.

264 "By that time . . . right": *Los Angeles Times*, 22 January 1984, p. D6.

265 Historian . . . 1970s.: John Walton, *Western Times and Water Wars: State Culture and Rebellion in California* (Berkeley: University of California Press, 1992), p. 231.

265 Violence returned . . . park.: *Los Angeles Times,* 30 August 1976, 16 and 17 September 1976, and 18 October 1987. See also Norris Hundley, *The Great Thirst: Californians and Water, 1770–1990s* (Berkeley: University of California Press, 1992), p. 343.

267 A dam at Eaton's . . . south: Irving Stone, "Desert Padre," *Saturday Evening Post* 216, no. 47 (1944): 9.

BIBLIOGRAPHY

I. ARCHIVAL MATERIALS AND MANUSCRIPTS

A. National Archives

Department of the Interior. Reclamation Service. Record Group 527. See the General File, 1902–1919 and Project File, 1902–1919.

B. Manuscript Collections

Austin, Mary Hunter. Papers. Special Collections, Henry E. Huntington Library, San Marino, Calif.

Early Water System Records. General Manager's Office Historical Records. Department of Water and Power, Los Angeles.

Hall, Pierson. Papers. Ventura County Museum of History and Art, Ventura, Calif.

Haynes, John Randolph. Papers. Department of Special Collections, University of California, Los Angeles.

Lippincott, Joseph B. Papers. Water Resources Center Archives, University of California, Berkeley.

McWilliams, Carey. Papers. Department of Special Collections, University of California, Los Angeles.

Mulholland, William. William Mulholland's Office Files. General Manager's Office Historical Records. Department of Water and Power, Los Angeles.

Outland, Charles. Papers. Ventura County Museum of History and Art, Ventura, Calif.

Owens Valley Historical Records. Real Estate Division Historical Records. Department of Water and Power, Los Angeles.

Sherman, Moses H. Papers. Sherman Foundation, Corona Del Mar, Calif.

Taylor, Raymond G. Papers. Los Angeles County Medical Association Library, Los Angeles.

C. Dissertations, Theses, and Miscellaneous Documents

Beneda, Janet. "The Los Angeles Aqueduct: The Men Who Constructed It." Thesis, institution unidentified, May 1974. Currently in the Los Angeles Department of Water and Power Library.

Cifarelli, Anthony. "The Owens River Aqueduct and the Los Angeles Times: A study in Early Twentieth-Century Business Ethics and Journalism." Master's thesis, University of California, Los Angeles, 1969.

Jones, William K. "The History of the Los Angeles Aqueduct." Thesis, University of Oklahoma, 1967.

Mulqueen, Stephen P. "The St. Francis Dam Failure," Ventura County Historical Museum.

Pentland, Gertrude. "Los Angeles Aqueduct with Special Reference to the Labor Problem." Thesis, Institution Unidentified, May 1916. Currently in the Los Angeles Department of Water and Power Library.

Spriggs, Elisabeth Mathieu. "The History of the Domestic Water Supply of Los Angeles." Master's thesis, University of Southern California, Los Angeles, 1931.

II. GOVERNMENT DOCUMENTS

A. United States

Department of the Interior. United States Geological Survey. *The Geology and Water Resources of Owens Valley California,* by Willis T. Lee. Water Supply Paper no. 181, 1906.

Department of the Interior. United States Geological Survey. *An Intensive Study of the Water Resources of a Part of the Owens Valley California,* by Charles H. Lee. Water Supply Paper no. 294, 1912.

Department of the Interior. United States Geological Survey. *The Quality of Surface Waters of California,* by Walton Van Winkle and Frederick M. Eaton. Water Supply Paper no. 237, 1910.

U.S. Bureau of the Census. U.S. 1900 Census, Los Angeles County, Calif. ED. 7, 61, 90, 123.

______. U.S. 1910 Census, Los Angeles County, Calif. ED 58, 123.

______. U.S. 1920 Census, Los Angeles County, Calif. ED 58, 255.

U.S. Congress. House. Committee on Irrigation and Reclamation. *Hearings . . . on Protection and Development of Lower Colorado River Basin: H.R. 2903, by Mr. Swing.* 68th Cong. 1st sess., 1924.

B. STATE OF CALIFORNIA

California Railroad Commission. *Los Angeles Aqueduct: General Construction and Auxiliary Costs,* compiled by O. E. Clemens, 1 February 1915.

Committee Report for the State of California. *Causes Leading to the Failure of the Saint Francis Dam.* Sacramento, 1928.

McClure, W. F. *Owens Valley–Los Angeles: Report Made at the Request of Governor Friend Wm. Richardson Following the Opening of the Alabama Hill Waste Gates of the Aqueduct by the People on November 16, 1924.* Sacramento, 1925.

Saint Francis Dam Commission. *Report of the Commission Appointed by Governor C. C. Young to Investigate Causes Leading to Failure of the Saint Francis Dam near Saugus California.* 1928.

C. CITY OF LOS ANGELES

Aqueduct Investigation Board. *Report of the Aqueduct Investigation Board to the City Council of Los Angeles.* 31 August 1912.

Board of Public Works. *Report of the Board of Consulting Engineers on the Project of the Los Angeles Aqueduct from Owens River to the San Fernando Valley,* by John R. Freeman, Frederick P. Stearns, and James D. Schuyler. 22 December 1906.

Board of Public Service Commissioners. *Twenty-second Annual Report.* 30 June 1923.

Board of Public Service Commissioners. *Report on Available Water Supply of City of Los Angeles and Metropolitan Area,* by Louis C. Hill, J. B. Lippincott, and A. L. Sonderegger. August 1924.

Board of Water Commissioners. Annual Reports.

City Council. *Report of the Committee to Investigate and Report the Cause of the Failure of the Saint Francis Dam.* March 1928.

City Council. *Report on the Los Angeles Aqueduct,* by Edward Johnson and Edward S. Cobb. 15 July 1912.

County of Los Angeles, County Coroner. *Transcript of Testimony and Verdict of the Coroner's Jury in the Inquest over Victims of the Saint Francis Dam Disaster.* Book 26902. April 1928.

Department of Public Service. *Complete Report on Construction of the Los Angeles Aqueduct.* 1916.

Department of Water and Power. Report of the Chief Engineer, Annual Reports.

District Attorney's Office. Report to Asa Keyes, *The Failure of the Saint Francis Dam,* by Allan E. Sedgwick, Louis Z. Johnson, Walter G. Clark, and Charles T. Leeds. March 1928.

D. Court Cases and Public Records

City of Los Angeles v. *Pomeroy*, 124 Cal. 597, 63 (1899)

City of Los Angeles v. *County of Inyo*, 167 Cal. App. 2d 736 (1959)

County of Inyo v. *City of Los Angeles*, 71 Cal. App. 3d 185 (1977).

County of Inyo v. *City of Los Angeles*, 78 Cal. App. 3d. 82 (1978).

Dornaleche v. *City of Los Angeles*, Case No. 273624

In the Matter of Estate and Guardianship of Lillian E. Sloan. 1919, Los Angeles County.

In Re Estate of Addie Haas Mulholland. Los Angeles County.

In Re Estate of Lillie Mulholland. Los Angeles County.

In Re Estate of Benjamin C. Strang. Los Angeles County.

H. H. Kelly v. *City of Los Angeles*, Case no. 273629

The Matter of the Estate of William Mulholland, File No. 151991, Los Angeles County.

The Matter of the Estate of Perry Mulholland, File No. 457423, Los Angeles County.

The Matter of the Estate of Clara S. Sloan. File No. 310299, Los Angeles County.

Ray E. Rising v. *City of Los Angeles*, Case no. 273625-628

Lucille M. Sloan v. *Edmund G. Sloan.* Divorce File, 1918, Los Angeles County.

III. PERIODICALS

A. Newspapers

Intake

Inyo Register

Los Angeles Daily News

Los Angeles Examiner

Los Angeles Express

Los Angeles Herald

Los Angeles Record

Los Angeles Star

Los Angeles Times

Los Angeles Tribune

Sacramento Union

B. Signed Articles

Brennecke, Olga. "How Los Angeles Built the Greatest Aqueduct in the World: A Story of Interesting Municipal Activity." *Craftsman*, November 1912: pp. 188–96.

Cross, Frederick C. "My Days on the Jawbone." *Westways*, May 1968, pp. 3–8.

Eaton, Henry G. "Poverty Revealed Our Riches." *Los Angeles Times Magazine,* 24 July 1932: 4.

Grimes, William. "Chunnel Vision." *New York Times Magazine*, 16 September 1990, pp. 34–37.

Griswold, Wesley. "The Day the Dam Burst." *Popular Science*, March 1964, pp. 88–92.

Grunsky, C. E. "Saint Francis Dam Failure." *Western Construction News*, 25 May, 25 June 1928, pp. 314–24.

Hall, Pierson M. "Review of the First Edition of *Man-made Disaster*." *California Historical Society Quarterly* 45, no. 1 (1964): 288–92.

Heinley, Burt A. "The Longest Aqueduct in the World." *Outlook*, 25 September 1909, pp. 215–20.

______. "Water for Millions." *Sunset* 23, no. 12 (1909): 631–38.

______. "Carrying Water Through a Desert." *National Geographic*, July 1910, pp. 568–96.

______. "Aladdin of the Aqueduct." *Sunset* 26, no. 4 (1912): 465–67.

______. "An Aqueduct Two Hundred and Forty Miles Long." *Scientific American*, 25 May 1912, p. 476.

______. "Restoring the Los Angeles Siphon." *Municipal Journal*, 7 May 1914, pp. 633–35.

______. "Aqueduct Outlet Cascades." *Engineering News*, 2 September 1915, pp. 455–56.

Heyser, Jack. "Los Angeles City Fathers Go Water Hunting: The Birth of the Owens River Aqueduct." *Journal of the West Antelope Valley Historical Society* 1, no. 1 (1988): 73–86.

Hoffman, Abraham. "Joseph Barlow Lippincott and the Owens Valley Controversy: Time for Revision." *Southern California Quarterly* 54 (Fall 1972): 239–54.

______. "Origins of a Controversy: The U.S. Reclamation Service and the Owens Valley–Los Angeles Water Dispute." *Arizona and the West* 19 (Winter 1977): 333–46.

______. "Did He or Didn't He? Fred Eaton's Role in the Owens Valley–Los Angeles Water Controversy." *Journal of the West* 22 (April 1983): 30–38.

Hurlburt, W. W. "William Mulholland, Man and Engineer." *Western Pipe and Steel News* 2, no. 1 (1925): 5–12.

Kahrl, William L. "The Politics of California Water." *California Historical Quarterly* 55, no. 1 (Spring 1976): 2–25.

______. "The Politics of California Water: Owens Valley and the Los Angeles Aqueduct, 1900–1927." *California Historical Quarterly* 55, no. 2 (Summer 1976): 98–120.

Lippincott, Joseph B. "Mulholland's Memory." *Civil Engineering* 9, no. 3 (1939): 199.

______. "Frederick Eaton." *American Society of Civil Engineers, Transactions* 100 (1935): 1645–47.

______. "William Mulholland: Engineer, Pioneer, Raconteur." *Civil Engineering* 2, no. 2 (1941): 105–7, 161–64.

Lissner, Meyer. "Bill Mulholland." *American Magazine* 73, no. 6 (1912): 674.

McCarthy, John Russell. "Water: The Story of Bill Mulholland." *Los*

Angeles Saturday Night, published in sixteen installments, 30 October 1937 - 26 March 1938.

Mulholland, William. "Water from the Colorado." *Community Builder*, March 1928, p. 23.

______. "A Brief Historical Sketch of the Growth of the Los Angeles City Water Department." *Public Service* 4, no. 6 (1920): 1–8.

Osborne, Henry Z. "The Completion of the Los Angeles Aqueduct." *Scientific American*, 8 November 1913, pp. 364–65.

Prosser, Richard. "William Mulholland, Maker of Los Angeles." *Western Construction News* 1, no. 8 (1926): 43–44.

Reed, Rochelle. "Castillo del Lago." *House and Garden*, May 1987, pp. 206–10.

Robinson, W. W. "Myth Making in the Los Angeles Area." *Southern California Quarterly*, March 1963, pp. 83–94.

Shrader, Roscoe E. "A Ditch in the Desert." *Scribner's*, May 1912, pp. 538–50.

Stewart, William R. "A Desert City's Far Reach for Water." *World's Work*, November 1907, pp. 9538–40.

Stone, Irving. "Desert Padre." *Saturday Evening Post* 216, no. 47 (1944): 9–11.

Twilegar, Burt I. "Mulholland's Pipe Dream." *Westways*, January 1949, pp. 16–17.

Van Norman, H. A. "Memoir of William Mulholland." *Transactions of the American Society of Civil Engineers* (1936): 1604–8.

Woehlke, Walter V. "The Rejuvenation of San Fernando." *Sunset* 28, no. 2 (1914): 357–66.

C. Unsigned Articles

"California's Little Civil War." *Literary Digest* 83, no. 10 (1924): 15.

"Dam Failure Wrecks Power Systems." *Electrical West*, 1 April 1928, p. 9.

"The Great Aqueduct and What It Means to Los Angeles." *Los Angeles Financier*, October 1910, pp. 3–7.

"H. A. Van Norman, Chief Engineer and General Manager, LADWP." *Intake* 21, no. 9 (1944): 1–16.

"Los Angeles Aqueduct." *Building and Engineering News*, 25 August 1915: 6.

"The Los Angeles Aqueduct Seizure—What Really Happened." *Fire and Water Engineering*, 17 December 1924, pp. 1312–13.

"The Owens Valley Controversy." *Outlook*, 13 July 1927, pp. 341–43.

"Nine Miles of Siphons," *Literary Digest* 46, no. 9 (1913): 452.

"Reconstruction of San Francisquito No. 2 Power Plant." *Electrical West* 62, no. 1 (1929): 14–17.

"Reconstruction of San Francisquito Power Plant No. 2 After Destruction by the Overwhelming Flood of March 13, 1928." *Los Angeles Section American Society of Civil Engineers* 2, no. 1 (1928): 11–16.

"Saint Francis Dam Disaster Invites Evil Politics." *Graphic*, 17 March 1928, p. 16.

"That Dreadful Night." *Westways*, March 1961, pp. 14–15.

D. Oral Histories

Department of Special Collections, University of California, Los Angeles

Department of Water and Power, Los Angeles

Eastern California Museum, Independence, Calif.

Ventura County Museum of History and Art, Ventura, Calif.

E. Miscellaneous

Clipping File, Los Angeles Municipal Reference Library, Los Angeles City Hall

Clipping File, Department of Water and Power, Los Angeles

Clipping File, Doheny Library, University of Southern California, Los Angeles

IV. BOOKS

Austin, Mary. *The Land of Little Rain*. Albuquerque: University of New Mexico Press, 1903.

______. *Earth Horizon: Autobiography*. Boston: Houghton Mifflin, 1932.

Bates, J. C., editor. *History of the Bench and Bar of California*. San Francisco: 1912.

Bonelli, William G. *Billion Dollar Blackjack: The Story of Corruption and the Los Angeles Times.* Beverly Hills, Calif.: Civic Research Press, 1954.

Campbell, Joseph. *The Power of Myth.* New York: Doubleday, 1988.

Chalfant, W. A. *The Story of Inyo.* Bishop, Calif.: Chalfant Press, 1922.

______, editor. *Constructive Californians.* Los Angeles: Saturday Night Publishing Co., 1926.

Davis, Mike. *City of Quartz.* New York: 1992.

Doyle, Helen MacKnight. *Mary Austin: Woman of Genius.* New York: Gotham House, 1939.

Ford, John Anson. *Thirty Explosive Years in Los Angeles County.* San Marino, Calif.: Huntington Library, 1961.

Gottlieb, Robert, and Irene Wolt. *Thinking Big: The Story of the Los Angeles Times, Its Publishers, and Their Influence on Southern California.* New York: G. P. Putnam's Sons, 1977.

Halberstam, David. *The Powers That Be.* New York: Knopf, 1979.

Hoffman, Abraham. *Vision or Villainy: Origins of the Owens Valley–Los Angeles Water Controversy.* College Station, Tex.: Texas A & M University Press, 1981.

Hundley, Norris, Jr. *The Great Thirst: Californians and Water, 1770–1990s.* Berkeley: University of California Press, 1992.

Kahrl, William L. *Water and Power: The Conflict Over Los Angeles' Water Supply in the Owens Valley.* Berkeley: University of California Press, 1982.

Kelly, Allen. *Pictorial History of the Aqueduct.* Los Angeles: *Times Mirror,* 1913.

McGroarty, John Steven. *Los Angeles: From the Mountains to the Sea.* Chicago: American Historical Society, 1921.

______. *History of Los Angeles County.* Chicago: American Historical Society, 1923.

McWilliams, Carey. *Southern California Country: An Island on the Land.* New York: Drell, Sloan & Pearce, 1946.

Matson, Robert William. *William Mulholland: A Forgotten Forefather.* Stockton, Calif.: Pacific Center for Western Studies, 1976.

Mayo, Morrow. *Los Angeles.* New York: Knopf, 1933.

Mulholland, Catherine. *The Owensmouth Baby: The Making of a San Fernando Valley Town.* Northridge, Calif.: Santa Susana Press, 1987.

Nadeau, Remi. *The Water Seekers.* Garden City, N.Y.: Doubleday, 1950.

Nordhoff, Charles. *California For Health, Pleasure, and Residence.* New York: Harper & Brothers, 1882.

Ostrom, Vincent. *Water and Politics: A Study of Water Policies and Administration in the Development of Los Angeles.* Los Angeles: Haynes Foundation, 1953.

Outland, Charles. *Man-made Disaster: The Story of the Saint Francis Dam.* Glendale, Calif.: Arthur H. Clarke Co., 1977.

Reisner, Mark. *Cadillac Desert: The American West and Its Disappearing Water.* New York: Viking, 1986.

Robinson, W. W. *Lawyers of Los Angeles: A History of the L.A. Bar Association and of the Bar of Los Angeles County.* Los Angeles: Los Angeles Bar Association, 1959.

Rogers, J. David. *Engineering Geology Practice in Southern California*, Belmont, Calif.: Star Publishing, 1992.

Soto, Gary, editor. *California Childhood: Recollections and Stories of the Golden State.* Berkeley: Creative Arts Book Company, 1988.

Starr, Kevin. *Material Dreams: Southern California Through the 1920s.* New York: Oxford University Press, 1990.

Sullivan, Mark. *Our Times: The United States 1900–1925.* New York: Scribner's Sons, 1926.

Taylor, Raymond. *Men, Medicine, and Water: The Building of the Los Angeles Aqueduct 1908–1913.* Los Angeles: Friends of LACMA Library, 1982.

Thomas, Lately. *The Vanishing Evangelist: The Aimee Semple McPherson Kidnapping Affair.* New York: Viking, 1959.

Walton, John. *Western Times and Water Wars: State Culture and Rebellion in California.* Berkeley: University of California Press, 1992.

Wood, Richard Coke. *The Owens Valley and the Los Angeles Water Controversy: Owens Valley as I Knew It.* Stockton, Calif.: Pacific Center for Western Historical Studies, 1973.

INDEX